Building Failures

Building Failures

Case Studies in Construction and Design

THOMAS H. McKAIG
Consulting Engineer

McGRAW-HILL BOOK COMPANY
New York Toronto London

BUILDING FAILURES

 Library of Congress Catalog Card Number: 62–8838

II

45107

Preface

This collection of case studies has been assembled to help architects, engineers, and contractors build more safely. There is no intention to point a finger at anyone. Most building failures are accidents, and those who are involved in them are the victims of accidents, to whom sympathy and commiseration should be extended.

At the same time, we are aware that accidents, too, are caused; and an examination of their causes is the first step toward their prevention. It is regrettable that the legal processes of determining liability for a failure so often serve to bury information of great potential value. The insurance carriers may be satisfied with a quiet settlement behind closed doors, but I strongly believe there is also a responsibility, in these cases, to the public.

Only if the reasons for a failure are understood and published have we who play a professional role in construction truly done all we can to prevent a recurrence. It is not too much to say that all our customary safeguards—building codes, standards, inspection procedures, safety factors—stem from unfavorable experiences. And we are still learning and improving our methods.

Experience can be a very expensive teacher, and fortunately we do not have to relive all the experiences of those who went before us. Men have been able to create such a vast and impressive technology precisely because of their ability to communicate their experiences—good and bad—and to build upon the recorded experiences of others. This book is intended as just such a record.

Thomas H. McKaig

Contents

1

This Book and Its Purpose

There usually is a reason for writing a book. It may be written to tell a story or to teach a lesson; or it may be that through the telling of a story, a lesson can be taught.

Building construction has been dotted with failures from the earliest days of history. Most of these failures have been forgotten, but because of them mankind has learned to build more safely. Each failure has carried a lesson for someone to learn, and because of these lessons structural design has reached its present stage of development.

It has been said that nobody learns anything from a building on which everything has gone smoothly. The earliest theories of structural design were developed by trial and error, mostly error. Therefore, one of the chapters in this book is "Lack of Precedent," and many other failures mentioned in this book could have been assigned that heading as of the date of the failure.

The primary purpose of this book is to tell the story of some structural failures for the benefit of those people whose livelihood may be affected by the lessons to be learned from them: architects, engineers, contractors, clerks-of-the-works, inspectors, building maintenance men, building owners, and others.

Although this book may interest the general reader, it is not intended primarily for him. The language used is that of the technical or semitechnical group mentioned above, but the textbook theory of some of the reports on these failures is not developed. Little attempt is made to argue the merits of controversial points; the desire is to present the facts as deduced from the press and from other available sources, such as technical magazines, daily newspapers, technical society reports, reports for insurance companies from the author's files, and decisions listed in legal abstracts. Obviously, only a small proportion of the failures that have occurred can be summarized, but an attempt has been made to select those failures that will be most significant to the modern practitioner or technician. The failure of a medieval cathedral roof in 1200 A.D. can have little educational value to the modern engineer, interesting as it may be historically.

1.1 Organization

In arranging the material in this book, it has been necessary to group the individual failures in chapters under general headings. The allocation of some of these failures has, of necessity, been arbitrary; many of them might have fitted into several of the categories listed.

It has been said that if a building shows no signs of distress before it is occupied it will continue to stand indefinitely. Although this statement is not strictly true, it is a fact that a much larger proportion of failures occur during construction than after a building is completed. Unless someone is killed or injured, many of these failures are not written up in the technical magazines; they may not even receive notice in the local papers.

These failures during construction have been arranged in groups based on the main cause of the failure and divided into chapters according to general nature: concrete, structural steel, miscellaneous failures (including masonry, timber, and the like), and failures during alterations. The failures discussed in Chap. 6–8—foundation failures, collapses attributed to carelessness or ignorance, and those caused by lack of precedent—may have occurred either during or after construction.

Some of the causes, such as the premature removal of concrete formwork and the overloading of floors, are covered many times on structures in different parts of the country. It is hoped that the large number of examples given may serve to emphasize the frequency and the tragedy of these errors of judgment. Thousands of lives have been sacrificed and tens of thousands have suffered injury, because someone was careless and, while trying to save a few dollars, speed construction, or get more load capacity out of a space, was willing to gamble with other people's lives.

The lack of detail on the cause of some of the failures is occasioned by the difficulty of learning the cause accurately. The word "accurately" is important, since a misstatement might unjustly point an accusing finger at an innocent person. All the facts available on many of the failures still leave discrepancies that preclude a positive statement of the real cause of the failure. For some of the collapses, on which several boards or committees have made investigations and reports, the reports differ greatly on the cause of the failure. For many of those failures that have been written up from magazine articles, the entire story is not available. Sometimes, when it was evident that litigation would take place, information was withheld or reporters were denied access to facts pending legal action, so that the engineering press had only the bare facts of the failure. Then, after the public—except those most closely involved—had forgotten about it, the case would be settled out of court, and all data that might have been of use to the engineering profession were forever buried.

Fire and storm damage, which constitute by far the most frequent bases of insurance claims, are passed over lightly because, for either of these categories, an entire volume could be written covering more failures than are considered in this book. Much of that material would be repetitious and of little value to those for whom this book is intended. A few of these cases are included, however, some in the chapters dealing with failures during construction; a few of the most famous are included in the chapter on fire and wind failures.

It should be carefully noted that the architect, engineer, contractor, inspector, plan examiner, and owner may all bear a legal responsibility for any negligence or acts of commission or omission that result either directly or indirectly in a failure, and that felony indictments are sometimes returned, and convictions obtained, in such cases.

It frequently happens that in the ensuing review of the failure, the investigators, the district attorney, or the coroner's jury urge legal measures to require better field inspection—responsibility of the architect or engineer for supervision as well as design—and more inspection by representatives of the municipal building department. In spite of all these warnings, however, an architect may still sell a set of plans and wash his hands of the responsibility for what the owner does with them. Building inspectors may still be appointed on the basis of whom they know at City Hall, or at party headquarters, rather than what they know. The technical man may have learned his lesson, but the politician and the penny-wise owner have not.

1.2 Why Buildings Fail

Usually, buildings fail through man's ignorance, carelessness, or greed.

Many years ago, so long ago that its source is not known to the author, a magazine published a list classifying "causes of failure" in construction. This list, which follows, need not be limited to buildings; it is broad enough to encompass bridges, docks, dams—any type of engineering construction. It is difficult to conceive of any failure or difficulty—major or minor—that does not fit into one of these classifications. The list is included here as it originally appeared; the author has noted in parentheses some building failures that might be assigned to each of these categories.

1. Ignorance
   a. Incompetent men in charge of design, construction, or inspection. (Sec. 11.2—"A potential concrete floor arch failure"—a contractor placed his bent bars in a concrete slab in the top of the slab instead of the bottom, expecting to get "arch action.")
   b. Supervision and maintenance by men without necessary intelligence. (Sec. 11.2—Kendall Bed Co.—a floor designed to carry 200 psf (pounds per square foot) actually loaded to 900 psf.)
   c. Assumption of vital responsibility by men without necessary intelligence. (Sec. 7.1—Knickerbocker Theater—a contributory cause of failure was the redesigning of the steel trusses by the steel company in order to use available material.)
   d. Competition without supervision. (Sec. 7.4—Atlanta Auditorium—the use of bar joists that did not comply with recognized standards.)
   e. Lack of precedent. (Sec. 8.1—Wilkins and Robins Air Force Base Warehouses—the failure of rigid frames designed by extremely competent engineers in accordance with best accepted standards of practice.)
   f. Lack of sufficient preliminary information. (Sec. 6.9—Ireland Building—the location of a footing so that one corner of it rested over an old filled-in cistern.)
2. Economy
   a. In first cost. (Sec. 11.1—Philadelphia General Hospital Food Service Building—omission of expansion joists as requested by the contractor and the architect.)
   b. In maintenance. (Sec. 11.1—"Leaning church steeples"—failure to provide properly against the tearing of flashing, thus allowing the wood at the base of steeple to rot.)
3. Lapses, or carelessness
   a. An engineer or architect, otherwise careful and competent, shows negligence in some certain part of the work. (Sec. 3.1—Brooklyn

Theater–lack of proper bracing to provide against possible erection stresses.)

b. A contractor or superintendent takes a chance, knowing he is taking it. (Sec. 2.1–New York Coliseum–lack of sufficient bracing in formwork.)

c. Lack of proper coordination in production of plans. (Sec. 4.2–"Bronx apartment collapse"–perhaps the precipitating cause of this failure, since it is evident that the designer's ideas were never interpreted to the contractor.)

4. Unusual occurrences

Earthquakes, extreme storms, fires, and the like. (All of Chap. 10.)

1.3 Historical Background

Even in Babylonian days, the hazards of the construction industry were recognized. In the Louvre in Paris, stands an 8-ft-high column on which the Code of Hammurabi (circa 1950 B.C.) is engraved. The Code is simple and direct:

If a contractor builds a house for a man, this man shall give the contractor two shekels of silver per ser [a unit of weight] as recompense.

If a contractor builds a house for a man and does not build it strong enough, and the house which he builds collapses and causes the death of the house owner, then the contractor shall be put to death.

If it causes the death of the son of the owner, then the son of the contractor shall be put to death.

If it causes the death of a slave of the owner, then he (the contractor) shall give the owner a slave of equal value.

If it destroys property, he (the contractor) shall replace what has been destroyed, and because he did not build the house strong enough and it collapsed, he shall rebuild the house at his own expense.

If a contractor builds a house for a man and does not build it so that it stands ordinary wear and a wall collapses, then he shall reinforce the wall at his own expense.

Apparently in Hammurabi's time there was little thought of unit stresses, live loads, and wall thicknesses; apparently, also, there was little chance of buck-passing. Responsibility was centralized in one man in the simple civilization of that day. Today the collapse may involve the contractor, the owner, the architect, the structural engineer, the building inspector, and the plan examiner of the building department.

As stated in the previous section, the failure of a medieval cathedral roof in 1200 A.D. can have little educational value to us today. Perhaps it would have much more if we knew all the facts relative to its construction: the foundations, the problems encountered when they were being laid, the nature of the failure, and the events leading up to it. Unfortunately, we do not, and we must therefore content ourselves with the story of the failures we do know about, with the types of construction we are living with today. Even the language of the construction industry of our day is different from that of other generations. If the story is to be read by engineers, architects, and builders of the twentieth century, it must be in the construction language of the twentieth century.

The author has chosen, therefore, to begin this story with the decade of the nineties, because that decade marks the practical beginning both of modern construction and of modern technical reporting. Steel had been made earlier, and the first skeleton frame building, the Home Insurance Building in Chicago, Ill., had been built in the eighties, but construction of this sort was experimental rather than typical. Wrought-iron beams had been used even before the Civil War, but they had been used with brick arches, not like the steel beam and concrete slab construction of today. Cast-iron columns in fire-resistant construction were used practically up to the turn of the century. During the nineties, concrete was used in experimental construction, usually in the form of arches rather than reinforced slabs. It was a decade of transition, and, therefore, an excellent starting point.

It is interesting to note that through the years construction work has become safer, judging by the diminishing proportion of building failures. Perhaps safety is being legislated into the industry through the requirements of building codes, through the right of building departments to enforce their ordinances, through the requirement of professional registration for designers, and through state labor laws that dictate safe practices.

Another safety factor is the establishment of quality standards for materials and workmanship, and for design practice, by the American Society for Testing Materials, the American Welding Society, the American Concrete Institute, the American Institute of Steel Construction, and other similar organizations; these standards are being written into specifications. Also, there is no doubt that the average construction man today has a better technical education than his counterpart a generation ago. Practically all contracting companies that undertake any sizable jobs today have graduate engineers in their organization, and it is not uncommon for some of them to be among the leaders of the engineering profession.

A sound economic reason also exists for the reduction in the number and size of construction failures. For workmen's compensation insurance, a contractor's premium rate is set by his safety record; thus there is no longer the incentive to take a chance that there once was, except in those lesser points that we might classify as lack of judgment.

Wherever the responsibility lies, it is certain that several times as many structures are built per year today as were built at the turn of the century, but fewer failures are reported (despite the improvement in news-gathering facilities).

It is interesting to note the variations in the types of failure that occur from decade to decade. Of course there are no abrupt dividing lines. Failures of old buildings from overloading and from willingness to take a chance will be noted throughout the entire 60 years covered by this book, and will no doubt continue to occur. (In fact, they will probably form an increasingly large proportion of failures as other causes of failure become better controlled.) Many of the collapses at the turn of the century, however, involved the failure of cast-iron columns, cast iron being brittle material that could not resist bending stresses such as those brought about by eccentricity of load. Cast-iron columns were then replaced by steel, a much more reliable material, and this type of failure ceased to occur.

The first decade of the twentieth century was the youth of reinforced-concrete design, and failures of that type of structure predominated. Many of those failures could probably be classified in the "Lack of Precedent" chapter, although the more obvious reason was the willingness of builders to take a chance and save money by either saving time or saving cement.

1.4 Elements of Structural Design

Although this book will most probably be read by men technically trained in building construction, some readers may not be familiar with the way that forces are resisted in a structure. The material in this section is intended to clarify, to some degree, the technical theory on which structural design is based. This theory is not difficult to understand; in fact, the fundamentals of structural theory are probably simpler than those of any other form of engineering and are demonstrated with such everyday items as cardboard boxes and rolled sheets of paper. This material is so elementary to technically trained men that they can pass over this section without any loss.

It is assumed that the reader is already familiar with those terms that are ordinarily used in the building trade: (1) a horizontal member that carries a load is a joist, beam, or girder, in ascending order of the area it carries (in fact, these terms are frequently interchangeable); (2) a vertical supporting member is a column—or, if it is a minor member, a strut; (3) a force tending to pull a fiber apart is a tension or tensile stress; and (4) a force tending to crush a fiber is a compression stress.

The principle illustrated in Fig. 1.1 is familiar to every student of high school physics—that of levers. If a bar is balanced over a fulcrum, a load of 25 lb at a distance of 8 ft from the fulcrum will balance a load of 100 lb at a distance of 2 ft. The product in each case of the load times the distance is called the moment: as long as the moment on one side of the fulcrum is equal to the moment on the other side, the bar is held in equilibrium, or balance, thus $25 \times 8 = 100 \times 2$ ft. A variation of this same principle is applicable to the design of a beam over two supports. The load on a simple beam is carried by the two supports. Now in accordance with Newton's Third Law, the mutual actions of any two bodies are always equal and in opposite directions along the same straight line—or, as

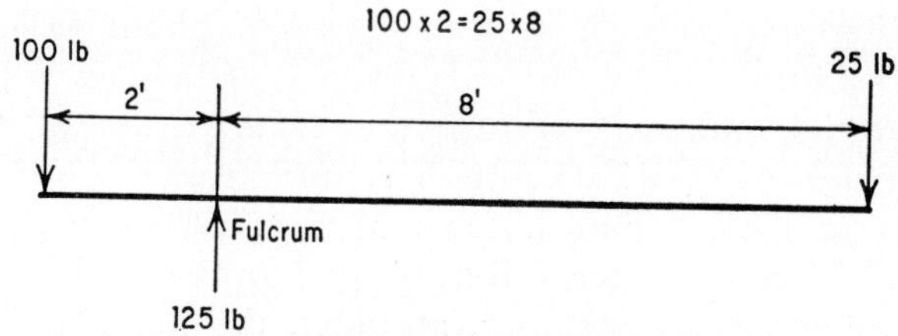

FIG. 1.1. Principle of levers

it is commonly stated, action and reaction are equal and opposite in direction. Using this concept, the resultant load under the end of a beam is spoken of as the reaction of the beam. If a member is to remain in a state of equilibrium, the algebraic sum of the loads on that member must equal zero; that is, the upward reaction must equal the total of the downward loads. Thus for the member shown in Fig. 1.1, the reaction upward from the fulcrum must be equal to the sum of the two downward loads—100 lb plus 25 lb, or 125 lb total.

Now, if the member in Fig. 1.1 is turned upside-down, so that a 125-lb load is placed on what was originally the fulcrum point, then the member has become a simple beam with a load of 125 lb and reactions of 25 lb and 100 lb. If the beam is to remain intact, if it is not to fail, then its internal strength must be sufficient to carry the stresses induced. These stresses can be demonstrated as follows:

If the reader will support an ordinary long rubber eraser at both ends and depress the center with his finger, he will have a demonstration of a beam bending (Fig. 1.2). In assuming a curve, the top fibers of the eraser shorten under compression and the bottom fibers lengthen under tension. This deformation varies from a maximum compression at the top face to a maximum tension at the bottom face: somewhere between the two is a plane that remains

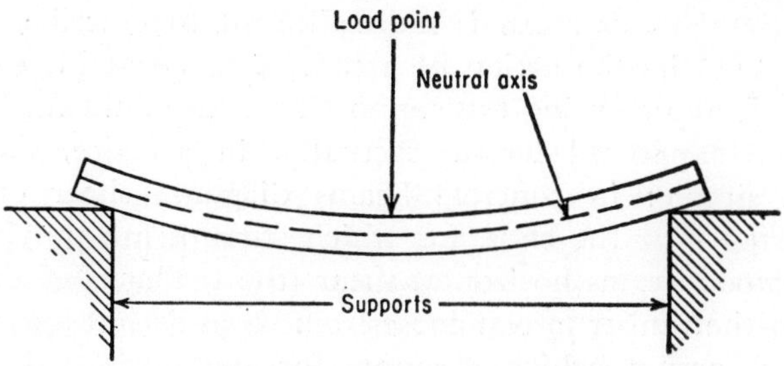

FIG. 1.2. Bending of a beam

its original length—the neutral axis. The internal stresses that result from bending in a homogeneous beam are indicated graphically in Fig. 1.3.

Briefly, then, the design of a beam requires the computation of the maximum moment the beam must carry. Then, with a knowledge

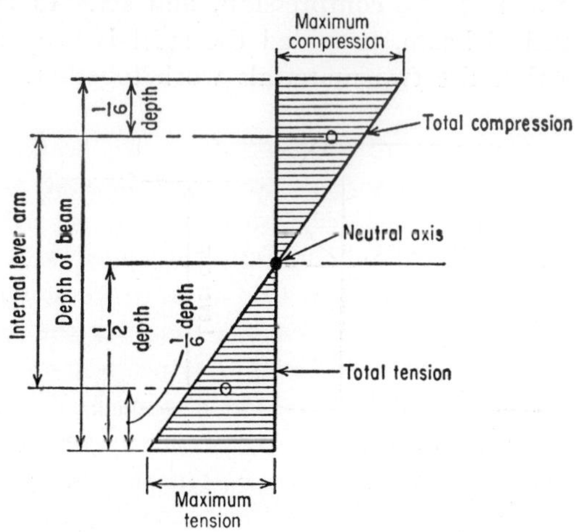

FIG. 1.3. Internal stresses developed in a homogeneous beam

of the maximum safe unit capacity of the beam material (the allowable unit stress), a beam section can be selected that will provide an internal moment sufficient to carry the maximum applied external moment without the maximum compression or tension stress (Fig. 1.3) exceeding the allowable unit stress for the material. The internal moment is the total compression or tension area (they are the same) multiplied by the lever arm between the center of gravity of each area (Fig. 1.3). This brief description can only begin to suggest the forces that work on a beam in bending.

In addition to tension and compression stresses, a shearing stress exists. Although this stress is more difficult to describe, it may be best understood by the action of a pair of shears, with the upward-moving blade representing the support and the downward blade the load on the supported beam. Actually, in practice, shear is not quite that simple: in concrete beams diagonal shear or diagonal tension occurs near the ends, for which stirrups are provided in the beams; in wood beams horizontal shear (the tendency of the horizontal fibers of the timber to tear loose from the adjacent fibers and slide along them) occurs, which accounts for the horizontal cracks frequently seen near the ends of a timber beam.

Figure 1.3, which indicates the stress distribution in a homogeneous section, such as a wood beam, is not applicable to a nonhomogeneous beam, such as a reinforced-concrete section. Concrete is able to resist compression stress much more efficiently per dollar spent than steel, but is weak and uncertain for tension. Steel, on the other hand, resists tension very well. Therefore, the two materials in combination—the concrete to resist compression, and steel to resist tension—make an economical beam, provided the steel is properly bonded to the concrete so that the two materials work together.

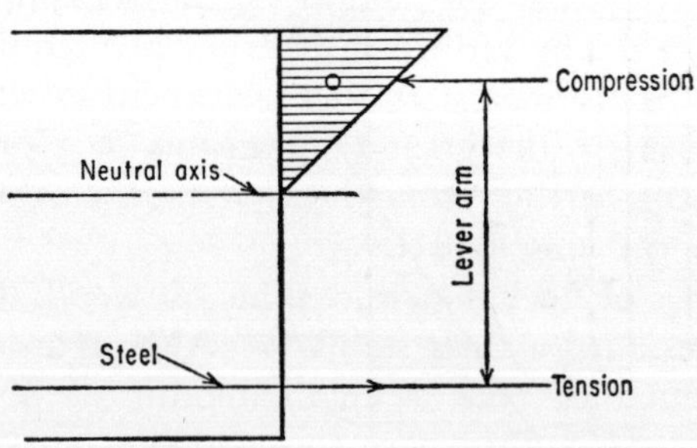

FIG. 1.4. Internal stresses developed in a reinforced-concrete beam

The internal stress diagram of a reinforced-concrete beam is shown in Fig. 1.4. Because of the different physical properties of the two materials, the neutral axis is not at the center, but approximately three-eighths of the way down the beam from the top surface. Thus, since the compression force acts at the center of gravity of the compression-force triangle—that is, one-third of the way from the top of the beam to the neutral axis—the lever arm for internal stress is approximately seven-eighths of the distance from the top surface of the concrete to the plane of the reinforcing steel.

Given a certain amount of material, the shape in which it is used will affect its ability to resist either bending as a beam or compression as a column, especially in steel members. For example: If the reader will place two books flat on a table about 9 in. apart to act as supports, then take an $8\frac{1}{2} \times 11$-in. sheet of paper and place it across these supports, the weight of the paper will cause it to sag. But the same sheet of paper rolled into a tube about 1 in. in diameter, held in that shape with rubber bands and placed on the same supports, will not only carry its own weight, but will, with another of these impromptu beams, carry a superimposed weight. The same sheet folded lengthwise in a series of V's or W's about $1\frac{1}{2}$ in. deep will also carry a superimposed load as a beam.

Similarly, as a column, the flat sheet of paper will not carry its own weight, nor can it be made to stand at all, but rolled into a tube it will stand on end and will carry a superimposed weight. To carry this analogy over into steel: a standard 3-in. pipe (outside diameter $3\frac{1}{2}$ in.) weighs 7.58 lb per ft; a $1\frac{11}{16}$-in. round bar weighs 7.6 lb per ft—practically the same. However, according to standard design theory, the pipe used as a column 7 ft high will safely carry a superimposed load of 36,050 lb, whereas the rod will only support a load of 13,900 lb. The capacity per square inch of a column depends on the slenderness ratio of its section—a relationship expressed mathematically as l/r, or the length of the section divided by the radius of gyration. The radius of gyration of a column depends on the distribution of its material: the further from the center of the section the material is, the greater the radius of gyration. Thus for any given area, the further from the center of a column its material is, the greater its carrying capacity.

Arches, whether of steel, concrete, stone, or wood, depend primarily on direct compression for their stability. Although steel and concrete arches are designed to take a small amount of bending, they are intended primarily for use where the load is comparatively uniform across the entire span. A load that weighs on only one side of an arch or a heavy concentration that throws the center of pressure outside the center third of the arch can cause the arch to fail.

Definitions

Some of the terms used in this book may not be familiar to all readers. A few of the more important ones follow.

Dead load: Dead load is the weight of the structure itself—the load that is always present.

Live load: Live load is the sum of all loads the structure must carry over and above its own weight—the weight of furniture, stored material, occupants, and the like.

Safety factor: The safety factor is the ratio of the strength of the material before failure to the safe working stress upon which the design was based. Although this factor can be less for certain materials, it is usually required to be two or more. In other words, under the full design load, most materials should not be stressed to more than half of what they should be able to carry.

Sidesway: Sidesway is used with the meaning generally applied by construction men—lack of sufficient stability against lateral forces—rather than the meaning it is given in the theory of design of rigid frames.

2

Concrete Failures

Although concrete is not a new material, the use of reinforced concrete, in which the steel takes the tension stresses and concrete the compressive stresses (Fig. 1.4), is comparatively new. Many of the early failures of reinforced concrete might have been included in Chap. 8, "Lack of Precedent," but are included here to compare what happened in the early years of the twentieth century with what is happening today. Compare, for example, the formwork failures listed here from the early years of the century down to the New York Coliseum. Is there any difference (except in the amount of detailed information we are able to get about the more recent failures)? [1]

2.1 Concrete Construction

Concrete was a job-manufactured material until recent years, when transit mixing became standard throughout most of the United States and Canada. During the years of job mixing, it was subject to all the human errors of judgment, carelessness, and ignorance. More has been learned about concrete in the past fifty years than any other construction material, and much of this knowledge has been obtained through the investigation of failures. In a 1903 report on a failure, the concrete was described in the investigator's language as "a good sloppy mix." Duff Abrams, in 1918, gave us the result of his tests, which show that "sloppiness" is harmful rather than desirable. His discoveries resulted in the general acceptance of the "water-cement ratio law," which states that in any plastic mix the strength and other desirable properties of concrete under any given job conditions vary directly with the ratio of cement to water used in the mix.

Today, with quality-controlled concrete, few failures can be traced to the quality of the material itself. The field work, however, is still subject to errors of judgment, carelessness, and ignorance: we still have formwork failures, frozen concrete, and concrete whose potential strength has not been attained because of human error.

In all states and territories of the U.S. the engineer who designs the structures must be a licensed professional engineer (or his counterpart, a registered architect). After the plans have been approved, however, their execution is entrusted, in most states, to a contractor who may or may not have the technical know-how to design such temporary items as formwork or to use safe measures in the construction of the building. Some states have taken the step of licensing contractors to insure professional competence. Sometimes adequate inspection is provided by the architect or engineer, but usually such inspection is so hedged in by the limitations of responsibility that the inspector or clerk-of-the-works hesitates to interfere with the contractor's procedure unless it is obviously unsafe. Otherwise the contractor may claim an extra for being required to do work beyond the scope of the specifications.

In the never-ending argument between the promoters of reinforced-concrete and those of steel-frame construction, time is always an element. In modern construction, concrete has answered the challenge of time by using large transit mixers and other equipment capable of placing 100 yd per hr, and motorized buggies that carry loads of a ton or more, equipped with brakes capable of stopping the load in a split second. All this equipment contributes new hazards to formwork construction; in spite of the advances in safety measures, concrete failures—particularly formwork failures—continue to occur.

New York Coliseum—New York, N.Y.

Probably the most publicized formwork failure in recent years occurred May 9, 1955, when almost 10,000 sq ft of the main exhibition floor of the New York Coliseum,[2] in New York, N.Y., collapsed, killing one man and injuring about fifty others. The floor that collapsed was the first floor above grade; at the time of the failure it was supported on two tiers of shores, a total of 22 ft. The lower tier consisted of dressed 4 × 4 wood shores spaced 2 ft 6 in. in one direction and 4 ft in the other. In the short direction they were capped with horizontal 4 × 4's; they were cross-braced in the other direction with horizontal 1⅛ × 9-in. boards. At mid-height the posts were braced in both directions with 1 × 4's. Directly over the 4 × 4 posts were adjustable metal shores carrying 4 × 6 girts; these girts in turn supported 2 × 5 joists spaced on 10- to 12-in. centers. A ⅝-in. plywood platform rested on the joists and carried the metal pans for the flat plate waffle slab (Fig. 2.1).

Fifteen bays of formwork had been erected, but pouring had been completed in only the first bay.

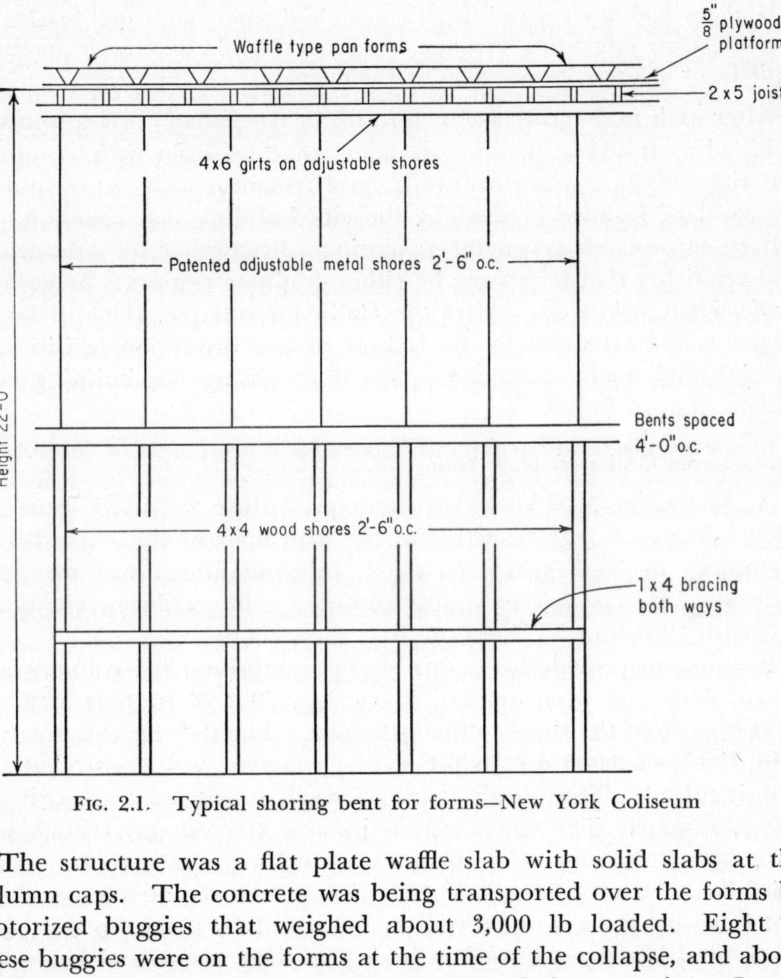

FIG. 2.1. Typical shoring bent for forms—New York Coliseum

The structure was a flat plate waffle slab with solid slabs at the column caps. The concrete was being transported over the forms by motorized buggies that weighed about 3,000 lb loaded. Eight of these buggies were on the forms at the time of the collapse, and about 500 cu yd of concrete had already been placed that morning. It was estimated that about 1,000 tons of concrete fell, carrying with it some of the forms from adjacent areas.

An investigation reported by the district attorney's office placed the blame on inadequate provisions in the formwork to resist lateral forces. The report stated that the failure was caused by:

> . . . inadequacy of the temporary supporting structure under the pressure of a horizontal or oblique thrust on the main exhibition floor. Such a thrust was created by the combined and sustained vibrations due to the movement and sudden stopping of the buggies, the dumping of the concrete, the operation of the concrete agitating machines, and the other activities on the floor.
>
> If there had been sufficient diagonal, horizontal, and end bracing of the temporary supporting structure, the collapse could have been prevented entirely, or at worst its area held to a minimum.

After each major collapse of formwork, particularly when someone is killed or injured, there are recommendations from various sources for tightening up on formwork requirements, such as: requiring approval of formwork plans by the city building department or by a licensed professional engineer; passing a license law for contractors; and requiring that inspectors be either graduate engineers or licensed professional engineers. After the Coliseum collapse, the district attorney called attention to the lack of proper inspection facilities on the job and made recommendations for revising the building code with respect to formwork.

Parke-Bernet Galleries—New York, N.Y.

A small-scale New-York-Coliseum-type failure occurred June 15, 1949, when a 39 × 76-ft section of the fifth floor of the Parke-Bernet Galleries [3] in New York, N.Y., fell, injuring fifteen workmen—two seriously. It was a 5-story-plus-penthouse reinforced-concrete building, with 19 ft 6 in. × 22 ft 6 in. bays throughout.

In order to provide for a third-floor auditorium, the columns and fourth floor had been omitted in an area 39 × 76 ft; that area was 22 ft high from the third to the fifth floor. The slab for this area was 7 in. thick, of 3,000 psi (lb per sq in.) concrete, well controlled and transit mixed. The forms, 5/8-in. plywood, rested on 2 × 4 stringers 2 ft on centers. The shores were adjustable at the top, 8 ft high, and were supported on 4 × 4 wood studs 14 ft long, which rested on the third floor 4 ft on centers. The shoring was braced at the top of the 4 × 4's with 4 × 4's in one direction and with 2-in. planks in the other. Horizontal bracing was used in both directions, above and below the 14-ft level. In all probability the criticisms made of the New York Coliseum shoring, a lack of diagonal shoring, could have been applied to the shoring used here.

During the morning of June 15, the concrete had been poured in three bays adjacent to the high shoring panels. After lunch, when the section over the auditorium was being poured, it suddenly gave way. It landed on the third floor, from which the shores had been removed, but the floor withstood the full impact of the blow. As a precautionary measure, however, the third floor was immediately reshored.

2.2 Other Formwork Failures

To avoid repetition, but to emphasize the frequency of occurrence of such failures, this section contains brief descriptions of a number of other formwork failures that are typical. More failures of this type do not make the headlines than do, and a number of other such collapses could be described.

Reservoir and filter plant roofs seem to be especially susceptible to this type of failure, probably because they are high single-story structures, and because their use does not dictate that great care be exercised to maintain them absolutely level as a floor slab. Also, although obviously formwork failures, sometimes it is difficult to point to the precise reason why the formwork failed. Each of the failures cited, however, emphasizes the necessity for care in design and installation of every part of the concrete forms.

Villa Scholastica College Chapel—Duluth, Minn.

On October 21, 1937, during construction of the Villa Scholastica College Chapel,[4] in Duluth, Minn., the formwork carrying the floor of a tower 28 × 30 ft collapsed while the concrete floor slab was being poured. The seven men who were working on the floor at the time fell 75 ft, and four of them were killed. Several salamanders (small portable stoves used to keep concrete from freezing) fell with the men and the concrete, and some of the men were burned as a result.

A preliminary investigation indicated that the falsework supports were inadequate for the load they were carrying.

University of California Mechanical Engineering Building—Berkeley, Calif.

On April 8, 1931, the roof of the University of California mechanical engineering building,[5] Berkeley, Calif., collapsed during the pouring of the concrete and fell 60 ft. Two men were killed and ten injured. The work had been inspected by a state inspector only three hours before the collapse occurred. After an investigation, the Industrial Accident Commission reported five possible causes of the failure as follows:

1. The patent locking device on one or more of the shores may have slipped, allowing the roof to sag.

2. The diagonal bracing of the shores may have been ineffective.

3. Some of the bracing may have accidentally been knocked out of place.

4. A possible thrust in the framework may have existed if the shoring was not plumb, or if it shifted due to the play in the hinged joint.

5. The bracing may have been insufficiently nailed.

Victoria Park Filtration Plant—Toronto, Ont.

In July 1934, while a 10-in.-thick roof slab was being poured for the Victoria Park Filtration Plant [6] at Toronto, Ont., the timber supports failed, dropping the men and freshly poured concrete 20 ft. Two men were killed and eleven injured.

Reservoir roof—Santa Monica, Calif.

In April 1949, 50 cu yd of concrete that had just been poured as the cover slab of a 5-million-gal underground reservoir[7] collapsed when the underpinning gave way and the forms failed. As a result of the accident, one man died and eleven were injured.

The theory of the collapse is that one of the double-tier shores was bumped by a wheelbarrow, thus starting a slow continuous failure until the entire roof was down.

Reservoir roof—Allentown, Pa.

On May 4, 1937, the final section of 18 roof sections of a new reinforced-concrete reservoir,[8] a 60 × 131-ft area, collapsed, apparently caused by the failure of a stringer under the steel pan form. About 75 cu yd of concrete had been poured over a 20 × 60-ft area when the forms collapsed under the weight of the wet concrete.

Failure occurred without any warning, beginning near the center of the section. The failure spread in a funnel-shaped manner. Inspection failed to show any indication of faulty formwork—either in workmanship or material—and it was impossible to determine which stringer failed first.

Filtration plant roof—Washington, D.C.

The flat slab roof of a 1,300-sq ft area of the Washington, D.C., filter plant [9] failed in August 1942, when the concreting was almost completed. Thirteen workmen were injured.

This was another double-tier formwork failure, generally similar to the New York Coliseum and to the Parke-Bernet Galleries. The actual triggering of the failure, however, was blamed on sabotage, but whether this accusation was based on fact or on war hysteria is not known.

Masonic Temple—Salina, Kan.

The *Engineering News-Record* of September 15, 1921 said, "So much was wrong with the design and the construction of the Masonic Temple [10] at Salina, Kan., which collapsed in July, that it would be easy to pass the failure legally by with the comment that there is no lesson to be learned from it." Nevertheless, any failure, the cause of which can be ascertained, may teach someone something, or at least emphasize by repetition a lesson that has already been taught. Although this example is classed as a form failure, it is not an example of the simple type of failure in which lack of lateral bracing is the sole cause of the collapse. Apparently there were other contributory causes including some rather complex ones.

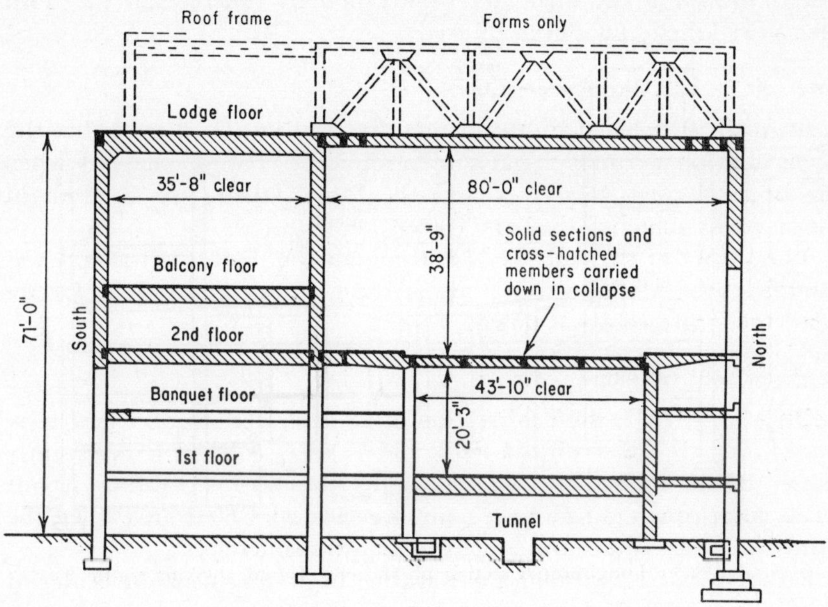

FIG. 2.2. Cross-section of Masonic Temple looking west

The building, which was approaching the "completion of the structure" stage, was a complicated layout of high and low floor heights as indicated in Fig. 2.2 and 2.3. Part of the area consisted of six floors but other portions, because of the layout, had only four stories in the same height. The requirements for lodge room, banquet hall, and auditorium introduced long clear spans (Fig. 2.4) and high stories which presented special problems.

One of these special problems was the use of two reinforced-concrete trusses over the top of the auditorium. Usually trusses are built of structural steel because concrete cannot carry the straight tension required in the lower chord of a truss; if the truss is made of concrete, steel reinforcing rods must still be used to carry this tension. Steel trusses are usually shop fabricated; 80-ft trusses such as were required for this building are ordinarily fabricated in two sections, then shipped to the job and spliced before erection. Thus, from the moment of their erection, steel trusses are able to carry load instead of depending on the floors to carry the load. In the case of reinforced concrete, the necessity of building from the bottom up required that the bottom

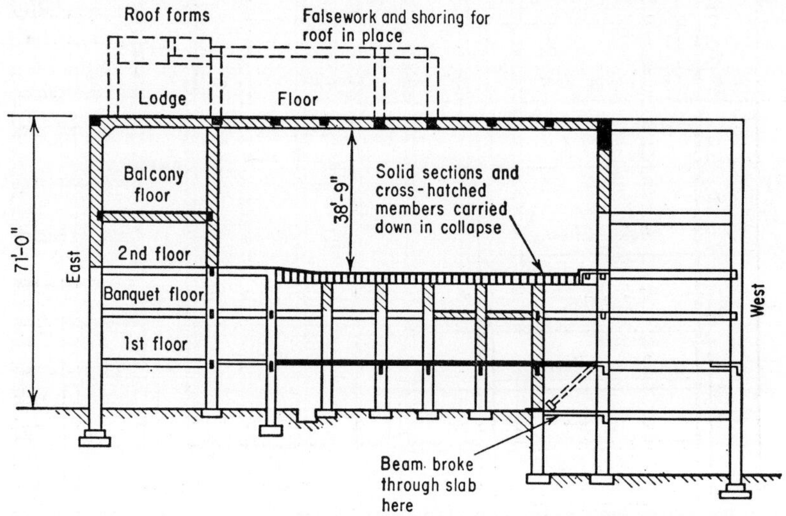

FIG. 2.3. Longitudinal section of Masonic Temple looking south

chord be poured as part of the ceiling of the auditorium, then the web members poured and finally the top chord. During this entire period, the weight of the truss is being carried by the floors below. Not until the entire truss assembly has acquired its full set, is it capable of carrying any of the load for which it was installed.

On July 11, 1921, when the collapse took place, the 90-ft-high reinforced-concrete framing was in place to an elevation of 71 ft above the ground; only the work of the top high story remained to be done. The concrete at the lodge floor level had been in place for five weeks, and the floors and columns beneath had been in place from eight weeks to five months. The formwork throughout, under the area where failure began, had been left in place, so no part of the responsibility could be placed on premature removal of forms. About an hour before the accident, workmen noticed some buckling of the shores and immediately started some emergency strengthening operations, which were not successful. A crash above them gave the men sufficient warning; everybody got out from under and no one was hurt, except two men who rode the wreckage down and received minor injuries. The collapse, however, destroyed about 80 per cent of the structure.

Investigations were immediately begun by a firm of consulting engineers; they found that the concrete had a low compressive strength —although no reason was given for this deficiency—and that the shores

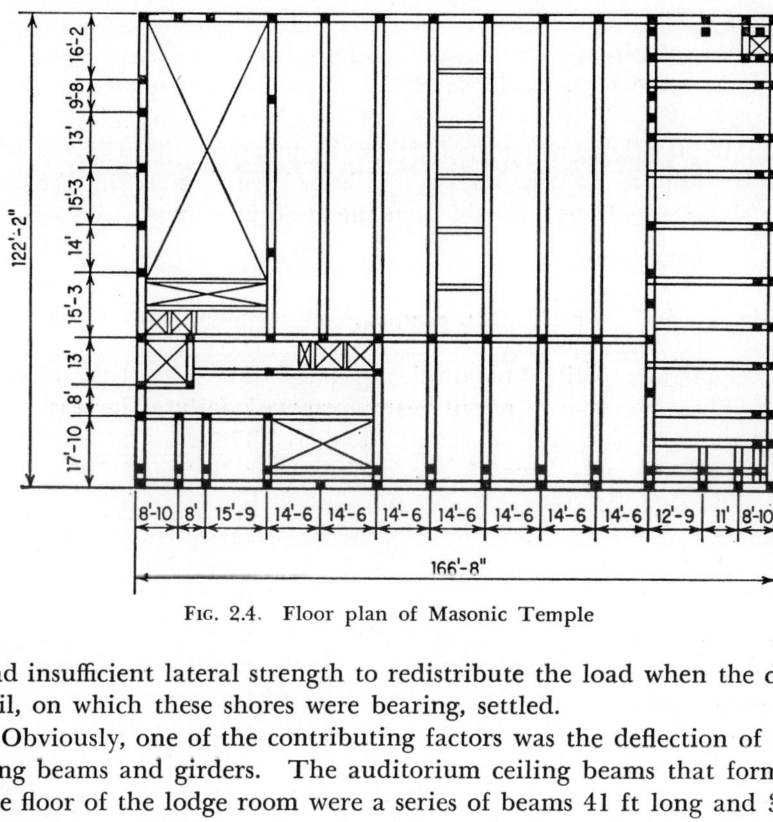

FIG. 2.4. Floor plan of Masonic Temple

had insufficient lateral strength to redistribute the load when the clay soil, on which these shores were bearing, settled.

Obviously, one of the contributing factors was the deflection of the long beams and girders. The auditorium ceiling beams that formed the floor of the lodge room were a series of beams 41 ft long and 3 ft on centers, and were supported on shores to the auditorium floor 37 ft below. These in turn were supported by a tier of shores 20 ft high to the first floor, which was shored to mud sills in the basement.

In parts of the structure, because of the variation in spans of supporting members and in the height of shores, the amount of deflection and lateral bending of the shores redistributed the loads so that some of the shores were loosened up, thus transferring further load. At the time of failure, which started in the center of the auditorium, it is difficult to know how much load any of the shores was carrying. The auditorium floor beams had a span of 43 ft 10 in. and were relatively shallow, so that when the mud sills settled, leaving the first floor without support, the weight of the entire dead load above came down on short beams, which were incapable of carrying this total dead load. A contributing cause of the failure was that some of the shores were carried on the concrete floor slab of the tunnel, whereas others were carried on the compressible earth adjacent to the tunnel.

Aside from the construction factors involved in the failure, one comment that applied to the design was that the structural features of the building pointed to a structural steel design, but that concrete had been substituted for the steel. This would apply primarily to the trusses, the high stories, and the long spans. The trusses have been discussed. The reference to high stories and long spans would apply in that the shoring, with all of its inherent difficulties, might have been omitted. The steel would have been erected and the concrete slab forms hung directly from the steel, an almost failure-proof method.

2.3 Three Recent Formwork Failures

For some unexplained reason, it appears that as other types of concrete failure decrease in recent years, formwork failures increase—not only in proportion but in actual number. Three such failures of major proportion occurred between November 1959 and June 1960. It is difficult to understand why this should be so, with greater experience, more of the work being done by specialty contractors, new mechanical safety gadgets for shores, better supervision, more safety consciousness induced by insurance rates, and the like, but these failures seem to increase in spite of these facts.

It is possible, of course, to speculate on the possible causes for this increase. With transit-mix concrete, it is common practice to pour concrete at such a rapid rate that the entire mass is still liquid, imposing its full hydraulic pressure on the formwork and shoring. With older, slower methods of pouring, the concrete at the bottom of the forms would have attained its initial set before the upper portion of the forms could be filled. Thus some of the fluid pressure would be supported by the hardening concrete itself. Moreover, the modern trend in concrete design is toward elimination of beams and girders and substitution of flat slabs, flat plate floors, and other types in which the formwork does not have the benefit of some beam action on the part of the forms or of the partially set concrete to carry or transfer any slight shift of the load that may occur.

Johnson and Johnson, Ltd.—Montreal, Que.

In the failure of a factory roof for Johnson and Johnson, Ltd.,[11] in Montreal, Que., in November 1959, a 120 × 160-ft area of concrete flat slab roof fell as workmen were placing the last 100 yd of a 750-cu yd pour. Forms and shores were said to have been substantially constructed to support the 9½-in.-thick slab on the 30 × 32-ft bays. The shores were 20 ft high.

A 45-mph wind was blowing, and it had rained heavily for several days before the failure occurred, but the builder discounted any probability of settlement of the mud sills due to this condition. At the time of the collapse the workmen were placing concrete on the deck from two truck cranes. The men rode the deck down when it failed, but no one on the deck was fatally injured. Four workmen who were pinned under the wreckage were killed, and a number of men were injured in the collapse. The formwork had been sublet to a "support and forms" company.

Municipal garage—Newark, N.J.

Another failure of the type mentioned in the preceding paragraphs was the collapse of an 18-in.-thick roof slab on a three-story underground municipal public parking garage[12] under construction in Newark, N.J., in March 1960.

The final crane load of a 300-cu yd pour had been placed and the workmen were screeding the deck when a rumble from below warned them of the collapse. Twenty men, who were working on the deck, scrambled to safety.

The roof was designed to carry the weight of a 4-ft earth fill for the park. A 2,000-sq ft area of concrete, weighting 240 tons, gave way and fell to the first floor below, where the slab was two weeks old. This slab also failed and the entire mass came to rest on the second floor below the roof, cracking the slab but not going any further. Despite the fact that the figures given appear large, the failure affected only 2 per cent of the entire floor area of the garage.

Inspection by building officials subsequent to the failure indicated that, when the forms for the floors below had been removed, they had not been sufficiently reshored to distribute the weight of the heavy pour into floors below, which had attained full strength to support the load. Evidence was found that some of the screw jack reshores appeared to have been bent before installation, and some of the reshores were not carefully plumbed.

Following the collapse, a four-month delay occurred during which investigations and tests were made, cores taken from the slabs, and load tests made. The slabs tested satisfactorily but core tests indicated that the concrete did not comply with specifications for mix proportions, and that it did not contain type III (high early strength) cement as specified.

Smithsonian Institution—Washington, D.C.

The last of this series of three formwork failures took place in June 1960, when an area of the first floor of the Smithsonian Institution [13] in Washington, D.C., fell, injuring twelve workmen. The area that fell—approximately 40 × 60 ft, weighing about 180 tons—fell about 25 ft into the basement. At the time of the failure, floor finishers were at work on the concrete of the first floor.

Several theories were advanced for the failure: (1) a faulty wedge; (2) a wedge knocked out of place; or (3) the possibility that some workmen in the basement had unknowingly bumped against a shore and pushed it out of place. Apparently one form joist started to drop and pulled the adjacent joists with it, bringing about a gradual chain reaction failure.

2.4 Lateral Failures of Forms

It is difficult to appreciate the high pressure that can be developed in the side forms of a concrete wall or deep beam under rapid pouring of concrete. Until the concrete has begun to set, the form is subjected to practically a liquid pressure—but the pressure of a liquid much heavier than water. Under rapid pouring, for example, the pressure on the bottom foot of a 6-ft-high concrete wall could develop to almost 900 psf, and to resist this pressure, heavy bracing is required—much heavier than one unaware of this pressure would normally furnish. The following two examples indicate this type of failure.

Dodge Engine Plant—Chicago, Ill.

Although the magazine article that reported this failure states that the "exact cause is unknown," the probability is that the type of lateral failure discussed above was the cause when the forms for the center wall in a divided drop-forge foundation pit at the Dodge Engine Plant, Chicago, Ill., failed, September 18, 1952, killing two carpenters and seriously injuring a third.

The men were at work in the pit when the substantially built form let go, trapping them in wet concrete and reinforcement. The wall was 16 ft wide and 20 ft high. The story is not complete on how much of the wall had been poured, but the fact that the men were trapped in the concrete and steel would indicate that most of the wall, if not all, had been poured. If we reconstruct what probably happened, we may develop this picture. A wall 16 ft wide and presumably 16 in thick would contain about $\frac{3}{4}$ cu yd of concrete per foot of height—or a 7-yd mixer load would fill the form to a depth of more than 9 ft. Fewer than three mixer loads would more than fill the forms, and this could all have been placed in less than an hour. According to a discussion of formwork pressures in the August 1958 issue of the *Journal of the American Concrete Institute,* the pressure could have been as much as 2,000 psf at the bottom form. It can readily be seen that, under the conditions described, even two truckloads of concrete could build up sufficient pressure to cause such a failure.

Vegetable market support beam—Los Angeles, Calif.

A formwork failure similar to that in the Dodge Engine Plant occurred when the formwork for a reinforced-concrete beam that was to be the main support beam for a large vegetable market roof [15] in Los Angeles, Calif., collapsed. The failure occurred September 23, 1940, killing one spectator and injuring several workmen.

The similarity in the two failures lies in the rapid pouring of concrete, and the building up of fluid pressure and vertical load before the concrete has had time to set sufficiently to partially sustain its own weight. The girder was 100 ft long, 16 ft deep, and 18 in. wide, and would have contained almost 90 cu yd of concrete on its completion, or about 120 tons. The pouring of transit-mix concrete into the beam forms began at 6 A.M. and continued until they were about two-thirds full at 3 P.M., at which time the formwork failed. Whether the side forms, as in the Dodge Engine Plant, or the supporting shores failed from lateral pressure was not explained in any of the reports of the failure.

Whatever the immediate cause of the failure, the accident emphasizes the need of: (1) adequate vertical support for the entire load; (2) adequate lateral ties or supports against spreading of forms; and (3) adequate bracing against sway caused by eccentricity.

2.5 Premature Removal of Forms

A frequent cause of failure during construction of concrete buildings is the premature removal of forms—either through a desire to obtain forms for reuse too soon, or because of delayed set due to cold weather. A group of typical failures which come under this category are described here, and in some of the sections following, the premature removal of forms was one of the major contributing causes.

The influence of freezing on concrete is discussed in Sec. 2.7. Even though the concrete has not been frozen, the delay of set due to cold weather can be the major cause in failure, because at temperatures below 40°F the chemical action of setting can be practically completely suspended. Even at temperatures below 50°F only a small proportion of the strength of concrete which has set at 70°F can be attained. Therefore, a month in late fall or early spring may produce less set in concrete than three or four days of summer weather. Temperature of curing, therefore, is very important in determining when forms may be removed. The following failures will emphasize this fact.

Shore Line Electric Railway—Saybrook, Conn.

The premature removal of forms, together with the overloading of a green concrete slab by piling up roof materials, caused the failure of three panels of a reinforced-concrete beam-and-slab roof of a new carbarn [16] for the Shore Line Electric Railway at Saybrook, Conn., April 7, 1910, resulting in the death of one man and injury to two others.

The roof structure consisted of 37-ft-long girders spaced 8 ft apart. The concrete had been in place for ten days when the shores were taken out, and, under the load of the roofing material, the end bay—the last bay poured—began to collapse. It is probable that without the added load, the failure may not have occurred. Although, under proper curing conditions, ten days should be long enough for setting of concrete today, it is probable that 50 years ago when job-mix concrete was used instead of transit-mix—when curing methods were not so well known—the ten-day removal was too short a time, even without the additional load.

Bridgman Brothers Co. Building—Philadelphia, Pa.

The misunderstanding of an order through language difficulties was the cause of the failure July 9, 1907 of part of a three-story reinforced-concrete factory building [17] being built in Philadelphia, Pa., for Bridgman Brothers Co. The failure resulted in the death of two men and injury to about thirty others. The building was a well-designed beam-and-girder-type building—in general, similar to the reinforced-concrete buildings of today. The concrete was a very rich mixture—1:1:2. (The method of specifying concrete in those days; that is, one part Portland cement, one part sand, and two parts broken stone or gravel. Today, concrete is specified by strength, such as 3,000-psi concrete.) The mixture used good quality sand and crushed trap rock or bluestone for the first two floors, and clean gravel for the third floor and roof.

The building regulations of Philadelphia set the requirement that shores must stay in place for a minimum of two weeks. The area of the roof in question had been poured for 5½ days when the subforeman in charge gave orders to a workman to remove alternate shores, and left this job for another job. The workman in turn transmitted the order to two Italian laborers, who misunderstood through their lack of knowledge of the English language, and proceeded to knock out all shores instead of alternate ones. How far they got is not known, but in all probability only a small area was needed to precipitate the collapse, and the continuous reinforcement pulled over the bays successively until an area two bays wide and five bays long—the entire length of the building—was in ruins. All of the floors went through to the basement except the first floor, where the failure was one bay wide.

Shanhouse and Sons—Rockford, Ill.

Although the use of flat-slab concrete construction is common practice nowadays and the method of design is part of any building code, this was not the case in the earlier years of its use. A number of patented systems were used then, any one of which produced the result of a flat slab—or what in those days was frequently referred to as "mushroom slab" (Fig. 2.5).

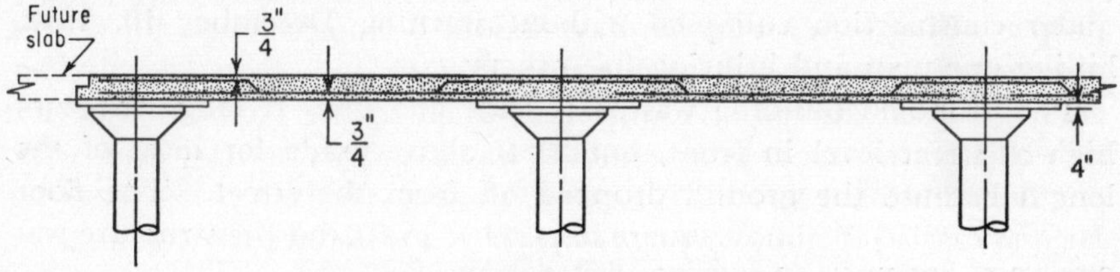

FIG. 2.5. Typical flat slab construction

On November 29, 1915, a floor of this type fell in a factory building [18] under construction for Shanhouse and Sons, in Rockford, Ill. Despite the fact that this is the only failure of this type cited in this chapter, the elements of the failure are remarkably similar to a number of others described here—improperly cured concrete, and premature removal of the shores—and the result too is the same, the building a total loss and two men killed.

The slabs were 7 in. thick with a 4-in. drop panel, a total of 11 in. thick at the column capital. The spans were 17 × 21 ft. The first floor slab was six weeks old, but in the interim the weather had been very cold. The concrete, although well proportioned, had a large amount of mixing water to make it flow readily around the reinforcing steel, a procedure that we now know gives a concrete of lower strength. Nowadays the result of better bond to the bars is usually achieved by mechanical vibration of the concrete, but in 1915, no such method was known, and the additional bond was accomplished by the use of a concrete which would flow readily.

The second floor slab had been poured on November 18, and had practically no set by the following day, when the men started stripping the first floor forms. The first indication of failure was a crack through one of the drop heads, and this was followed by a failure in shear around the column caps.

Several investigations were begun immediately, and the findings were the same in the various reports—indications of frost crystals around the aggregate, segregation of the heavy aggregate as a result of too much water, and a weak porous concrete from frost action and delayed set. Samples tested at the University of Illinois showed concrete of a low strength. The report of the coroner's jury placed the blame on premature removal of the supports, and recommended that the city adopt a building ordinance to cover concrete construction, and employ a qualified building inspector to enforce it.

Eastern Avenue moving picture theater—Cincinnati, Ohio

The first reinforced-concrete collapse of this type of building marred the safety record of Cincinnati, Ohio, when a motion picture theater [19] under construction collapsed without warning December 10, 1912, killing one man and injuring five others.

The proposed building was 31 ft wide and 98½ ft long, one-story high at street level in front, but 25 ft above grade for most of the length because the ground dropped off from the street. The floor slab was carried in almost square bays, 14 × 15 ft, and the structure was figured for 80-lb live load.

The official plans were drawn by an architect and a capable engineer, but the owner, to save money, gave the job to a contractor who drew his own plans, based on the original drawings. This contractor, a carpenter by trade, built the forms and sublet the concrete work. Neither of them had any previous experience with reinforced-concrete work and the contractor figured on reusing his form boards for roof sheathing. There was no competent supervision of any kind on the job. The brick walls were built on the concrete before any forming was removed. Following this, all slab forms and side forms of the beams were removed for use in the roof construction, leaving the beam bottoms in place. Since this still did not supply enough lumber to finish the roof, the beam bottoms were removed, and the collapse took place.

Just one month had elapsed from the time of pouring of the floor to the failure, and during this time the temperature was below freezing on several days, and below normal most of the time. In general, the construction was in conformance with the plans, and there is a possibility that, if the concrete had been properly cured before the shores were removed, the building would have stood despite the inexperience of those who built it.

2.6 Removal of Forms as a Contributing Cause of Failure

Several of the failures that have occurred when forms were removed were probably the result of other causes, and one could scarcely blame the premature removal of forms. Other contributing causes might have brought about failure even though the forms had been left in place for several months. The following are examples of this type of failure, even though the report at the time of the accident was "premature removal of forms."

Bixby Hotel—Long Beach, Calif.

The Bixby Hotel [20] at Long Beach, Calif., a five-story concrete structure, was nearing the completion of the pouring stage when, November 9, 1906, a section of the roof being poured fell—carrying the entire section through to the basement and killing ten men. The lower floors were of tile-and-joist construction, and the roof was a 4-in.-thick slab. The roof was shored to the fifth floor, which had been placed 18 days earlier.

Investigation indicated that the foundations and all other construction in the building were sound; therefore, the failure was attributed to "premature removal" of shores supporting the fifth floor and continuation of construction before the concrete was properly set to support the superimposed load. Later computations for the concrete columns, however, indicated that in view of the theory of design of that time, the concrete columns were seriously underdesigned. With a recommended allowable stress of 300 psi, compression stresses in the columns of this building ran as high as 923 psi, of which the dead load constituted more than half. F. E. Turneaure and E. R. Maurer's *Principles of Reinforced-Concrete Construction* (John Wiley & Sons, Inc., 1909 edition) recommends an allowable stress of 400 to 450 psi. Even today, with all the progress of more than fifty years of concrete manufacture and know-how, the allowable stress for 2,000-psi concrete is 450 psi.

Because the building had been designed according to the best knowledge of the time—with the one exception described above—this failure attracted more comment in the technical press from those who were really concrete specialists than anything up to that time, and thereby probably contributed much to the knowledge of concrete design as of that date.

Eastman Kodak Co.—Rochester, N.Y.

A collapse that attracted a great deal of publicity occurred November 21, 1906, when a 53 × 80-ft portion of a new reinforced-concrete building[21] for Eastman Kodak Co. at Rochester, N.Y., collapsed, resulting in the death of four men and the injury of several others. The building was a tile-and-concrete-joist structure designed and built by Concrete Steel and Tile Construction Co. of Detroit, Mich.

Investigation of the collapse indicated many evidences of defective workmanship. Many of the columns showed a horizontal joint of sawdust, shavings, and paper—in some cases several inches deep. In one place, a piece of wood 5½ in. wide and 13 in. long showed on the face of a concrete column. In other places the reinforcement showed on the surface of the beams. In several places the concrete was of poor quality, and forms had bulged from lack of proper form bracing. Strangely enough, despite the shock of the fall, the first floor did not fail.

The beam steel did not lap over the columns in many places, and, in the portions that fell, steel did not lap from story to story in the columns, nor were dowels provided.

A report made by independent experts employed by the owner stated that "the primary cause of the failure was due to the fact that the supports under girders and floor construction were removed too soon for that season of the year"; the record showed that the concrete was only about three weeks old. Later the owner made several floor load tests, which indicated that the floors could safely carry from 3.84 to 7 times the design load, or based on dead and live load, a safety factor of 2.54 and 4.09, respectively, for the two tests—both without failures.

Stark-Lyman Building—Cedar Rapids, Iowa

The reports of the failure of the Stark-Lyman Building[22] at Cedar Rapids, Iowa, November 14, 1913, is—as far as the failures of that period go—largely a repetition of others. One end of a seven-story building with 20 × 20 bays (Fig. 2.6), and tile-and-joist floors, 2 bays deep, and 3 bays wide, collapsed (Fig. 2.7), killing seven men. All the floors and part of the roof were in place at the time of failure. Shores had been removed to the underside of the fifth floor, leaving the sixth, seventh, and roof shores in place. The design had been

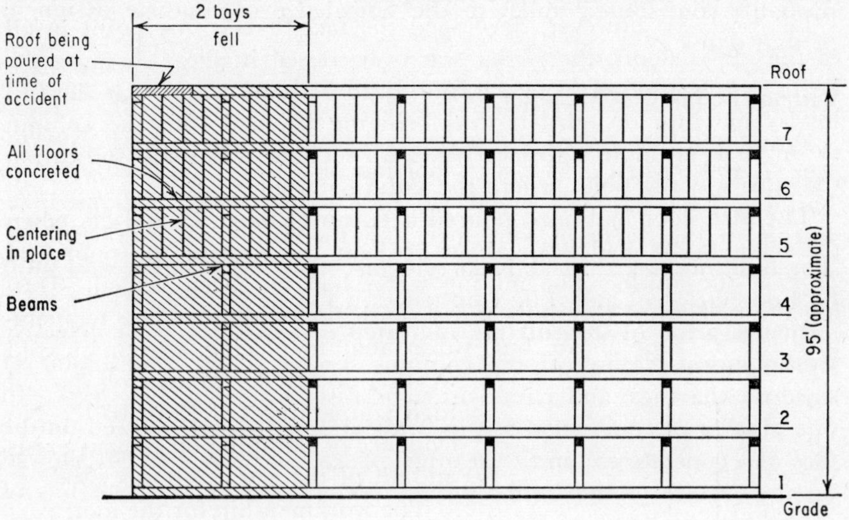

FIG. 2.6. Elevation of Stark-Lyman Building showing portion that collapsed

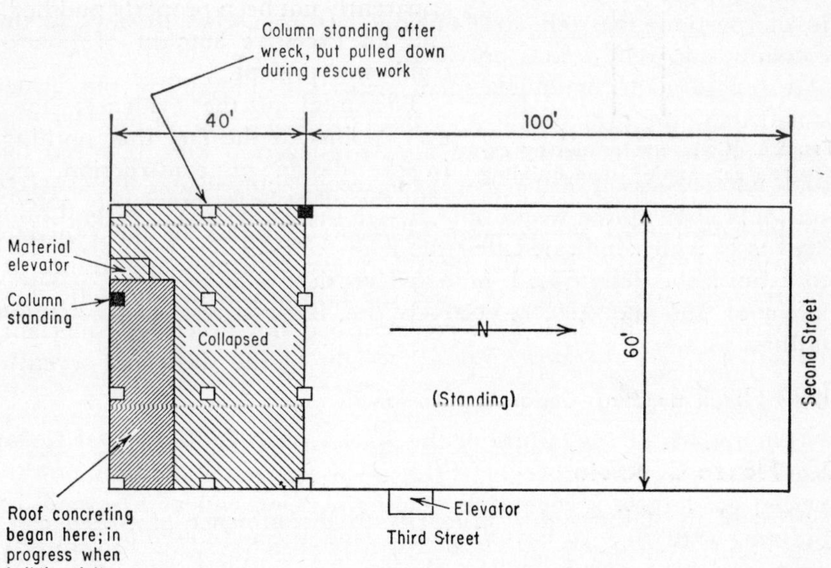

FIG. 2.7. Plan of Stark-Lyman Building

sent to a prominent authority on reinforced concrete to be checked, and he approved the plans as submitted.

Later review, however, developed the fact that at one point in all of the upper floors, the girder was overstressed in shear, because the beam had been offset from the column to such an extent that the full load came into the girder, thus bringing its load into the column through shear in the girder (Fig. 2.8). There was some indication that this was the point where the failure started.

The fifth floor had been poured between October 13 and 20, when the temperature had been about 50°F at the start of the pour and 21°F at the finish. The mixing water had not been heated. During the placing of the sixth and seventh floors, the daytime temperature was 43°F or above, but, subsequent to placing, extremely cold weather occurred. Many days showed high temperatures of 12 to 16°F. No evidence of skimping on cement exists and all of the aggregate was good.

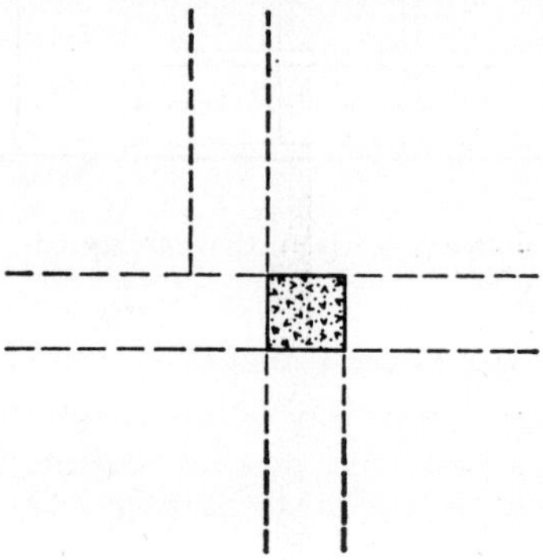

FIG. 2.8. Condition producing excessive shear—Stark-Lyman Building

The workmanship for the most part was good, but in places the alignment of tile was such that the width of joist varied. The upper columns had apparently not been properly puddled, and an excessive amount of honeycomb was present.

The total of the evidence, however, points to the fact that nothing in the design or construction, except the offset beam previously noted, would have caused the collapse by itself. Samples of the concrete following the failure had the typical "green concrete" appearance of concrete that has not fully set, and the failure was undoubtedly due to the fact that the fifth floor concrete had not developed sufficient strength to carry the superimposed weight of the sixth and seventh floors and the roof.

2.7 Frozen Concrete

Most of the failures due primarily to the presence of frozen concrete occurred in the decades from 1910 to 1930, the young adulthood of this type of construction. Concrete had been accepted as a good sound construction material, the general theory of design had been developed, and, in general, modern methods of construction had been or were being adopted. However, as has so frequently been stated in connection with construction failures, familiarity breeds contempt. Certain practices had been worked out, proven to be satisfactory, and accepted as being standard, and precautionary measures were often disregarded. In an effort to spread construction work over a greater part of the year, magazines urged that concreting operations continue throughout the year, and listed methods of protection for winter concreting. Although the failures cited in these articles will show that, in general, some precautionary measures were taken, those precautions were insufficient.

Although some facts were known about the dangers of frozen concrete, they were not as well known as they are now. We know today that concrete sets very slowly at temperatures below 50°F, and practically not at all at temperatures below 40°F. Tests have shown that concrete exposed to a temperature of 16°F, after a period of six hours curing at 70°F, has a 28-day compressive strength of less than 10 per cent of the same concrete moist-cured at 70°F for twenty-eight days. Several cycles of freezing and thawing in the early stages of setting are more injurious to concrete than continuous freezing. Even if the materials of the mix and the forms on which it is placed are properly heated to 70°F, the protection and curing of the concrete during the first week after placing is as important as the heating itself.

To the best of the author's knowledge, there has been no major failure caused by frozen concrete for the past thirty years, although undoubtedly some have occurred in which frozen concrete was a contributing factor. The reason for this is not definitely known, but it is certain that the man on the job today is more conscious of the hazards than was his predecessor; undoubtedly inspection is better; and, in many instances, methods of operation are dictated by building codes. Heated ready-mix concrete can be bought, plastic enclosure materials are available, and the contractor may use heaters which are far superior to yesterday's salamanders.

Apartment house—Buffalo, N.Y.

Typical of the frozen-concrete type of collapse was the failure of a single bay on each of two successive days of a twelve-story apartment house in Buffalo, N.Y. One bay failed December 28, 1927; the corresponding bay on the opposite side of the building failed December 29, 1927.

The building was a steel skeleton structure, with steelform concrete joists. The two wings were identical and projected out one bay beyond the balance of the building so that the joists were semicontinuous, rods being made continuous with the adjacent slabs on the inside end only. There was nothing unusual about the job; the plans were checked by the Buffalo Building Department and the permit issued without any changes.

The foundations were placed in early August and early September; the structural frame was erected very rapidly; and the concrete floors were placed about a week to ten days apart. In the upper floors as colder weather approached, this schedule was lengthened somewhat. On December 10, when the eighth floor was poured, the temperature in the morning was 59°F, but it dropped rapidly through the day, so that in the evening it was down to 12°F. Tarpaulins had been placed to enclose the area, and salamanders were lighted to provide heat for setting the concrete. When the pour was finished late in the afternoon, the temperature was about 20°F. During the evening a gale started to blow off Lake Erie. The weather report for the day showed a wind velocity of 86 mph, with a 3.8-in. snowfall. About midnight, when the wind was at its height, the tarpaulins were ripped off and the salamanders and hot coals inside were blown from the seventh floor to the street below. The fire department was called out and stood by to extinguish any live coals. One of the tarpaulins was retrieved about three blocks from the site. The day after this gale, the temperature was 17°F below normal (the average of high and low for the day), reaching a maximum of 19°F. This below-normal temperature continued for two days, then six days of above-normal temperatures occurred, then ten days on which below-normal temperatures were registered on all except one day. During this period there was very little work done.

The salamanders were not rekindled for three or four days and possibly not at all, but work was resumed and forming placed for the ninth floor. The eighth floor shores were still in place on the seventh floor, and perhaps part of the shores were in place for several stories below this. On the 27th, a thaw began; the snow 5-in. deep; on the 28th, 3-in.; and on the 29th, 1-in. The temperature was well above normal, 13°F above on the 28th, and 17°F above on the 29th.

The work progressed rapidly on the morning of the 28th. The brick walls on the bays that failed were finished up to the seventh floor level, and brick was piled on the seventh floor for the bricklayers. The ninth floor had been partially poured the day before and was finished by 10 A.M. To allow the laborers to get through to the perimeter of the seventh floor with wheelbarrow loads of bricks and to pile these bricks for the bricklayers, it was necessary to remove some of the shores in the eighth floor. This was done by 3 to 3:30 P.M., and ten minutes later the southwest bay collapsed, killing several men. The eighth floor failure took out all the floors below it to the basement. Strangely, the ninth floor did not fall. The next morning, before the shores could be replaced, the same bay on the southwest wing failed taking everything out from the ninth floor down.

Nobody was in this part of the building and there were no injuries.

Numerous frost crystals were found around the aggregate in the eighth floor concrete that failed, and it could be crumpled readily by hand. The building department ordered floor-load tests to be made of the seventh, eighth, and ninth floors. The seventh and ninth floors tested very well—far above the requirement—but the eighth floor slabs deflected so badly in two places that the building department required corrective measures; the floor was gunited (sprayed with mortar) and retested with very satisfactory results.

Hotel Vincent—Benton Harbor, Mich.

It must have been weird to watch an eight-story hotel practically melt away into a pile of broken concrete, gravel, and reinforcement, over a thirty-hour period, from January 28 to January 30, 1924. The reinforced-concrete Hotel Vincent[23] at Benton Harbor, Mich., had been started in late November 1923, and two months later, while the eighth floor was being poured, the failure began. Throughout the entire period the construction had been carried on at top speed, regardless of the weather. Some concessions had been made to temperatures in that aggregates and mixing water had been heated, an oil burner had been placed in the mixer, and calcium chloride had been used in the mix, but no attempt was made to check the amount of set which had been attained. A few salamanders had been used and protection by means of tarpaulins suspended to enclose the floor below each pour had been used but had been moved up with each successive floor, so that the concrete had a maximum of less than a week of protection, such as it was, throughout the bitter cold of January 1924.

The first two floors of the building, to a roof at third-floor level, were 100 ft square with a 56-ft square court in the back corner for the upper six floors. The reinforced-concrete construction of the upper floors was of concrete joists, using metal pan forms, and a slab known in building terms as a 6 plus 2 slab, with 5-in.-wide joists, 25 in. on centers. This joist construction was carried on beams parallel with the street fronts and spaced about 14 ft on centers. The design checked to carry not only the dead load to which it was subjected at the time of the failure, but a normal live load as well. Properly constructed, there was no reason why failure should have occurred.

On Monday, January 28, part of the eighth floor had been poured, and the shores were removed from under the fifth floor. All floors above this point were shored, and all shores had previously been removed from the lower floors. The fifth floor had been poured on December 28 and 29 when the minimum temperature was 24°F, and this was followed by two severe cold waves, with temperatures going down to 2°F and 16°F, respectively. Two days after finishing the concreting of the fifth floor, the concreting of the sixth floor began. Since tarpaulins were available for enclosing only one story at a time, the 2-in.-thick fifth floor slab had a maximum of two days protection. The fourth-floor columns and fifth-floor slab were exposed to near zero temperatures less than 48 hours after they were completed. The work was proceeding rapidly, and the seventh- and eighth-floor slabs were poured just before below-zero temperatures. As has been noted, the forms for the fifth floor were removed on January 28, replacing partial shoring under the beams. At midnight of January 28, the three west-central bays of the fifth floor began to sag, together with all the floors above. By 9 A.M. on January 29, a gash three bays wide and one bay deep for the full height of the building appeared down one side of the structure. Before noon it had extended to two bays deep. Shortly afterwards the columns began to fail on the other side of the building and an area the entire seven bays in length and one bay deep from the street collapsed. No further failure occurred during the afternoon, but after nightfall the progressive failure continued, and by 9 A.M. on the 30th, the entire structure was a heap of ruins with the exception of the two-story-high wing under the court—all of which had been finished before the extremely cold weather set in. The wreckage, for the most part, had disintegrated to the point where it gave the appearance of a pile of trash to be shoveled away, rather than pulled apart by means of a crane. It was with great difficulty that pieces could be found big enough to permit any proper crushing test to be made.

It was evident from the first that this was purely and entirely a cold-weather concrete failure. Beams, joists, and columns were properly designed and reinforcement properly placed. There were evidences of improper mix—not enough cement, too much sand, and a possibility of a cement which may not have complied in all respects with ASTM requirements—but none of these, without the contributing factor of cold weather and improper cold weather precautions, would have brought about the collapse themselves. Very few records of temperature, pours, and the like, were available. No cement tests or cylinder tests were made. The Portland cement was "Michigan Portland cement," furnished by a local dealer. The fine aggregate was a fairly good, clean quality of sand. The crushed stone was graduated in size from fine to coarse, and had a fairly large amount of crusher dust in it, but not enough to disqualify it for use. The specifications called for a 1:2:4 concrete. A quick rule of thumb would call for about 1.5 barrels of cement per yard of concrete for this mix, which would roughly furnish 2,000-psi concrete. Although the amount of cement delivered would not quite give this quality of concrete, it was not far short.

It is in the matter of cold weather precautions, however, that the real trouble lay. In 1923 and 1924 the proper treatment of concrete in cold weather was not too well known. We have criteria today, however, by which to measure the relative safety of the methods used and, only by comparing the methods they used with today's standard accepted methods, can we indicate the real cause of the failure.

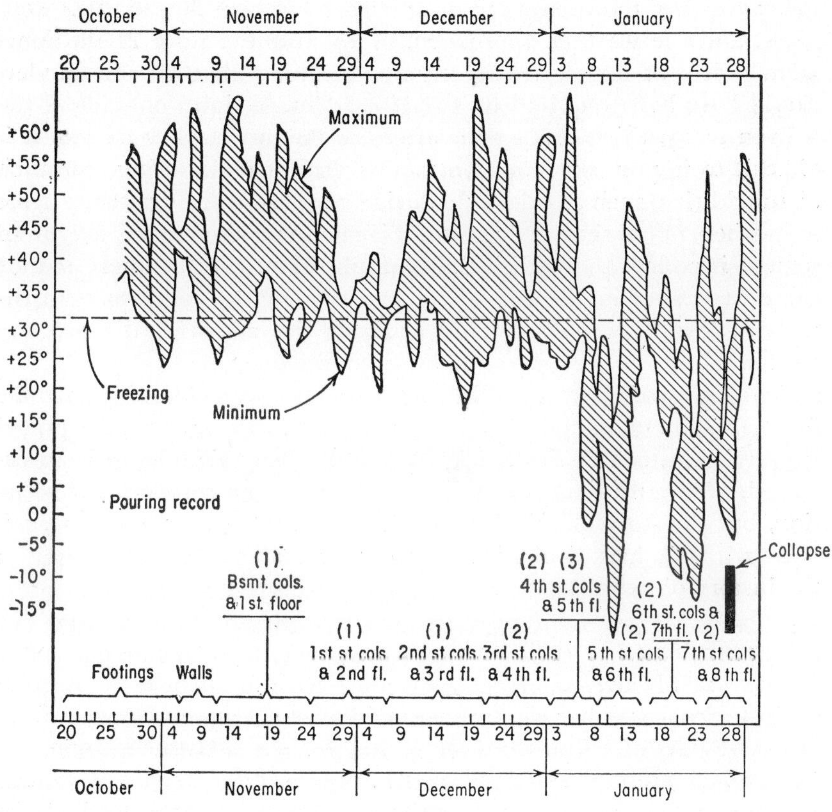

FIG. 2.9. Temperature record during construction of Hotel Vincent

The temperatures during the pouring period, November, December, and January (Fig. 2.9), were below 40°F every day except two. During January, after the pouring of the fifth floor, every day except one, at the same time of day, showed a temperature of 30°F. On eleven of these January days, the temperature dropped to a low of zero or below. During this period the mixing water and aggregates were heated, presumably satisfactorily, and an oil burner was used to heat the mixture. We can only assume that the material, as it left the mixer, was satisfactory. We have no knowledge of whether the metal forms were heated before the concrete was placed.

The top surface of the 2-in.-thick slab was exposed to freezing temperature within a matter of hours after placing, so it may be assumed that part of this slab was frozen before it had attained practically any set.

Sixteen salamanders were used on the entire job. Good practice today requires the use of one salamander for each 300 sq ft of floor space, and the weather approaching zero would require 2 additional salamanders at each outside column, meaning that 79 salamanders would have been required on the Hotel Vincent job.

Even today, thirty-seven years after the Benton Harbor failure, it is difficult to impress on some contractors the danger of using too much of an additive, such as calcium chloride. Most authorities today agree that when more than 1 per cent of calcium chloride by weight of cement is added, the additive is of doubtful value, and more than 2 per cent will definitely reduce the strength of the concrete. Apparently at least 4 per cent calcium chloride was added to the concrete on this job.

Five pieces of concrete satisfactory for tests were found, and these five tests showed strengths of 425 psi, 718 psi, 456 psi, and 258 psi respectively, instead of the 1,500 to 2,000 psi that might have been expected. Since no casualties occurred, there was no coroner's investigation, but the story of this failure probably did more to impress engineers with the hazard of winter concreting than any other collapse in the history of concrete.

John Evans Hotel—Evanston, Ill.

The use of dirty aggregate, together with insufficient protection against winter weather conditions combined to bring about the failure of twelve bays of a four-story reinforced-concrete apartment hotel, the John Evans Hotel[24] at Evanston, Ill., November 3, 1925. Workmen were removing the forms under the first floor while the load of the roof and all floors above were shored to the first floor. The impending failure gave sufficient warning so that all workmen had time to escape.

The work had been progressing very rapidly and no records had been kept of the pouring of the concrete. The contractor estimated that the first floor, which was of flat-slab construction, had been in place for 20 to 28 days at the time of the failure. Although it was claimed that the concrete on the upper floors was protected by salamanders and tarpaulins for several days after each pour, photographs taken after the collapse showed no evidence of them. The official average of the temperatures for twenty-five days before the failure was 41.1°F, just above the 40°F temperature considered to be the minimum at which any set takes place and, of these 25 days, 3 days showed average temperatures below 32°F and 10 were below 40°F.

The ground floor was an 8½-in.-thick flat slab, designed for 100-psf live load, and the upper floors were of metal-formed concrete joists designed for 40-psf live load. The dead load of the upper floors and roof was about 60 psf, or a total of 240 psf above the first floor. Even assuming that the ground floor had attained 70 per cent of its set and the upper floors 40 per cent average, this would place about 144 psf on a slab capable of carrying less than this load without the aid of the first floor shores.

Not only were the slabs insufficiently protected, but sand from the excavation had been used for all the upper floors, sand that showed from 8 to 10 per cent clay and silt, thus slowing the set and lowering the strength of the concrete. No cylinder tests were made on the concrete, and, since nobody was killed or injured, no official investigation report of the failure was given to the public.

Spang Chalfant Building—Etna, Pa.

The hazards of winter construction work, with several assists from questionable practice, are emphasized in the failure, February 19, 1912, of a portion of a three-story beam-and-girder-type reinforced-concrete building [25] for the Spang Chalfant Co. at Etna, Pa., a suburb of Pittsburgh. The *Engineering News* for February 29, 1912, gives two versions of the failure—one by Mr. Edward Godfrey, a consulting engineer of national reputation, the other by the architect and designer.

The roof of the building had been placed several days before Christmas, and several weeks later a cinder fill was placed to furnish pitch—the fill running up to a depth of 26 in. During the period after its placing, it had been saturated and frozen so that to all intents and purposes, at the time of the failure, it was practically solid ice up to 2 ft thick. From the time the roof slab was poured, up to the time of the collapse, the weather had been below freezing most of the time—several times below zero—and little had been done to put heat into the building. As long as a concrete slab remains frozen solid, it will have the strength it requires. When the ice thaws and becomes water, the only resistance to compression is a very weak concrete—in this case, a concrete that could be crumbled by hand. Most of the shoring had been removed some time before the collapse, but a heavy thaw set in on February 16, reducing the ice to water and precipitating the collapse under the heavy load of saturated fill (Fig. 2.10–2.13).

Mr. Godfrey also considered the probability of a foundation failure

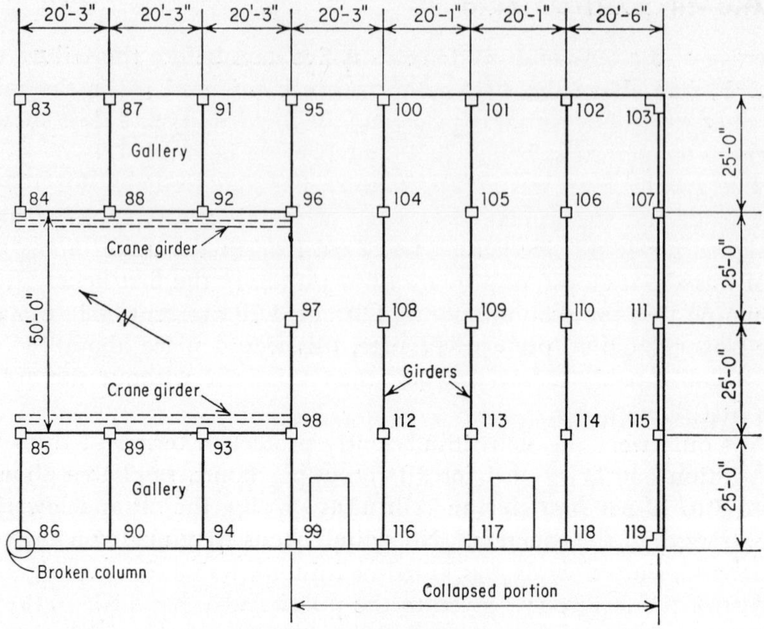

FIG. 2.10. Plan of Spang Chalfant Building

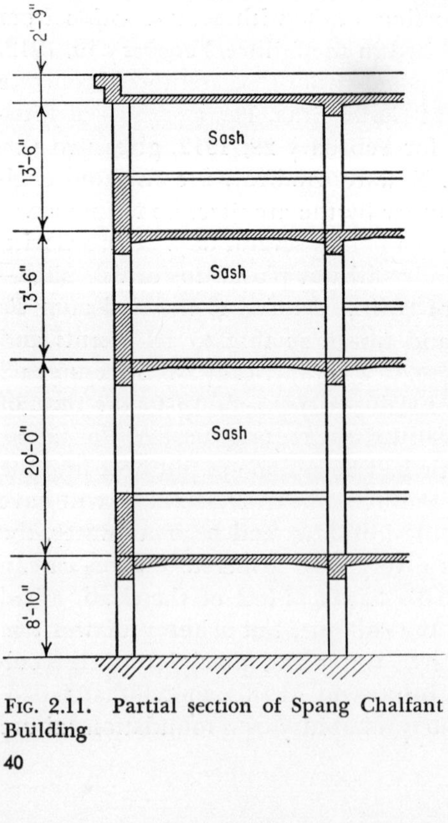

FIG. 2.11. Partial section of Spang Chalfant Building

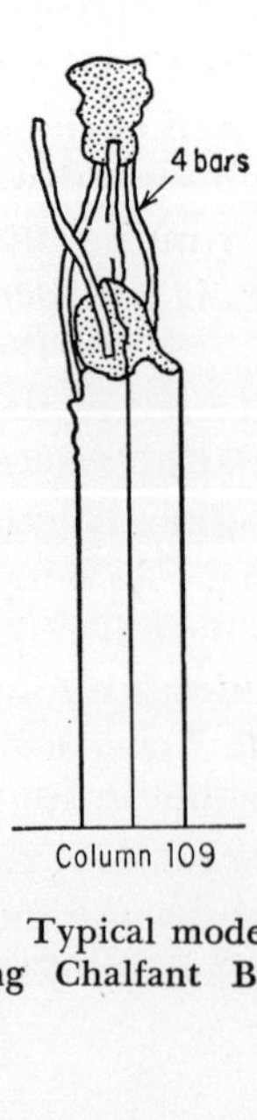

FIG. 2.12. Typical mode of failure—Spang Chalfant Building

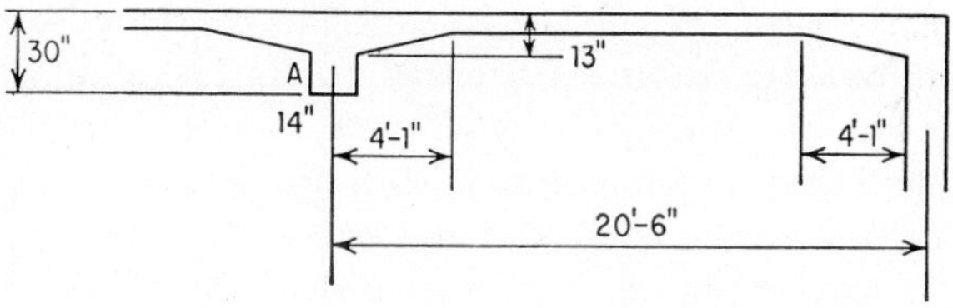

FIG. 2.13. Floor slab detail of Spang Chalfant Building

caused by overloading one column through a change in plans after the foundation of this column had been placed.

2.8 Lack of Precedent

Although many of the early failures of reinforced-concrete structures might have been included in Chap. 8, "Lack of Precedent," the cases listed in this section might better be listed under the topic of carelessness.

A material had been developed that was easily handled, could be placed by common labor, and could be made in the field. Those who became familiar with its various uses became careless; those who were not familiar with it assumed they knew enough about it to build with it. Some early mistakes with this material are listed in this section.

Reed's Bath House—Atlantic City, N.J.

In March 1906, only a small amount of newspaper publicity was given to a reinforced bath house [26] failure in Atlantic City, N.J. A report by an engineer correspondent for *Engineering News,* however, yields the following: There was much indication of frozen concrete and spalling (surface chipping and crumbling) from this cause. Moreover, the floor beams had apparently been poured to the underside of the slab and the slab poured in a separate operation, thus furnishing a much less effective depth than had been intended by the designer. Bearing walls poorly laid up with improper bond and very poor mortar were contributing causes to the failure.

Wing and Bostwick Building—Corning, N.Y.

On the morning of December 15, 1903, a section of a practically completed department store building [27] for Wing and Bostwick at Corning, N.Y., failed (Fig. 2.14). No deaths or injuries resulted. The portion that failed consisted of the two rear bays of the roof and the rear bay of the three floors. Five bays of reinforced-concrete construction were used altogether. Inspection of the section that failed showed that reinforcement had been omitted at several critical points.

In those early years of reinforced-concrete construction, when the theory of reinforced-concrete design was still in its infancy, practically no architects knew anything about it. Even among engineers, a comparatively small number had much skill in design. The American Concrete Institute (ACI) Code did not exist, and most of the design work was left to the engineers who furnished the reinforcement.

The contractor admitted that he knew nothing about concrete design, and, at that time, it was common practice for the architect to

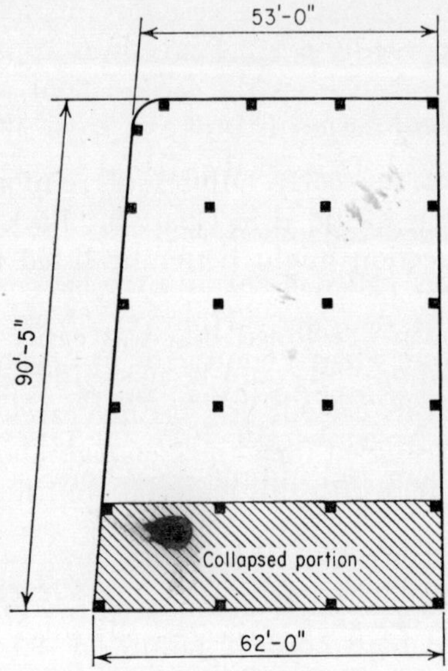

FIG. 2.14. Plan of Wing and Bostwick Building

write into the specifications the live load for which the design was to be made.

In this instance, the firm that sold the steel to the contractor and furnished the engineering services had primarily been in the business of making fences up to that time, and had only recently gone into the steel-reinforcement business.

Cowhey Building—Detroit, Mich.

The failure of the Cowhey Building [28] in Detroit, Mich., November 15, 1912, causing the death of three men and injury to a number of others, combined most of the faults of the worst days of reinforced-concrete construction as described in preceding cases.

The design was made by an architect with no experience in reinforced-concrete construction; his drawings were turned over to a local steel company for the reinforcing design, and he did not supervise the job. Instead, the work supervision was in the hands of a superintendent for the owner who resigned before the job was done, leaving the hoisting engineer in charge of the job. The second floor had been in place for a month, the shores had been removed from under it, and the third floor was being placed when the collapse occurred, involving 3½ square bays. The floor was a tile-and-joist floor, carried on tee beams in bays 17 ft 2 in. × 17 ft 10 in. The building department issued the building permit without examining the plans, giving the excuse that they did not have the necessary manpower.

The concrete, supposedly a 1:2:4 mix, was apparently quite weak. Month-old concrete taken from the second floor could be broken by hand. Tested in compression it broke at from 400 to 500 psi, much less than the theoretical stress on the slab at the time of the failure.

Concrete building failure—Logansport, Ind.

Many structures are called upon to carry loads during construction undreamed of by the designer. However, there is such a thing as carrying it too far, and, when a contractor in Logansport, Ind., hoisted his concrete mixer up to the third floor for the purpose of completing the concrete coping, November 10, 1914, the falsework failed.[29] The collapse killed one man and injured six others.

Henke Building—Cleveland, Ohio

The failure of the Henke Building[30] in Cleveland, Ohio, on the evening of November 22, 1910 was especially tragic because one of the walls fell on a two-story frame building next door, crushing the building, killing four of its occupants, and injuring seven others (Fig. 2.15).

The building replaced a building that had been destroyed by fire the previous May. A portion of the wall of the old building was left standing and was incorporated in the new structure. This portion, however, did not fail in the collapse. The new building had been designed as a timber-mill construction building with brick walls, but later reinforced-concrete construction was substituted for the timber-mill type. The type adopted was of standard reinforced-concrete column-and-girder construction, with a tile-and-joist concrete floor slab and exterior bearing walls. It was to have been a four-story-and-roof building 90 × 103 ft in plan. The contract called for a very tight time schedule—a completion date for enclosure of building by November 10—90 working days after the date of the contract.

This, however, was not accomplished; at the date of the collapse the roof had only been poured twelve days, and shores were still in place under the roof and under the fourth floor, as well as some of the shoring under the third floor girders.

The cause of the failure was not learned immediately, and an investigating committee was appointed, representing the Builders Exchange, the Cleveland Engineering Society and the American Institute of Architects. The whole structure had fallen straight down so that no part of any floor was more than 2 or 3 ft horizontally from the position it had occupied when in place. Many of the columns were still standing more than two stories high, and the floors had stripped from the columns.

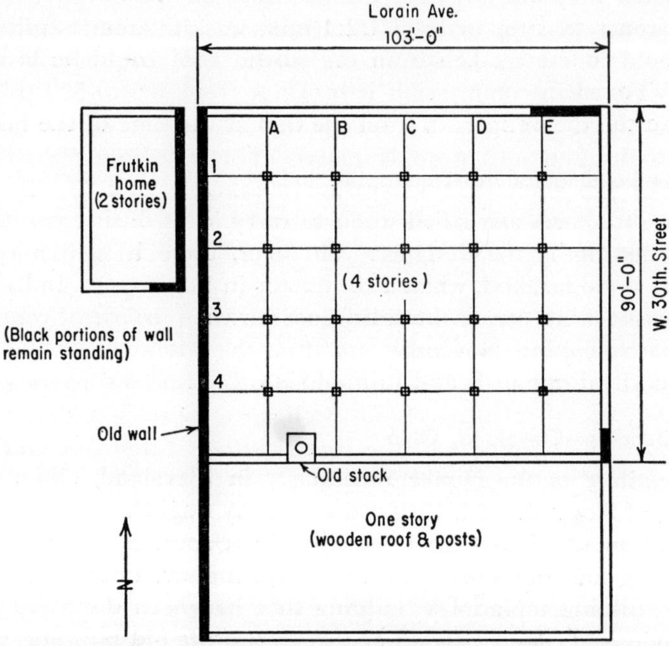

FIG. 2.15. Plan of Henke Building

The report of the investigating committee pointed the responsibility toward: (1) the architects—not because of design but because of lax specifications and failure in administration of the duties required of them by the Cleveland Building Code; (2) the contractors—for general negligence in construction in a number of items; (3) the owner—for failure to employ a concrete inspector as advised by the architect; and (4) the building department, for failing to enforce a portion of the code requiring that a concrete inspector be named before the permit was issued.

A section of the Cleveland Building Code required that whenever concrete work was used, the owner must name in writing a concrete inspector who would be at the work continually during the mixing and placing of concrete and reinforcement; this inspector must make daily reports to the building department. The architect had advised the owner to do so, but had not told him that it was mandatory. The building department had issued the permit without this notification in writing and had never protested the fact that daily reports were not being filed.

The report of the investigators cited the following evidences of negligence on the part of the contractor:

1. Inexperienced and totally incompetent foremen in charge of the work.

2. Careless placing of steel reinforcement (in some cases column ties spaced as far as 42 in. apart); girder rods improperly and insecurely placed.

3. Structural members of concrete not of proper size; concrete of poor quality and not properly placed (3-in.-wide joists between tile instead of 4-in., honeycomb, sawdust in concrete, and the like).

4. Dirty and inferior quality of sand used.

5. A shortage of 20 to 25 per cent in the amount of cement used as checked from delivery slips.

6. Removal of forms without authorization of the architect as required by the building code.

A report published in the *Engineering News* for February 23 gives a number of other instances of negligence, bearing out, and even going beyond, the report of the investigators. Lumps of clay up to the size of a hen's egg were found in the concrete. Sawdust was found in layers in the columns and mixed with the concrete. One column bar was split longitudinally—apparently a bar of rerolled rail steel. Column bars that were supposed to have been butted together end-to-end and held in place with sleeves were not cut square, and the sleeves were sometimes so large that instead of butting they were alongside each other. The brick face and backup were not bonded, but a face brick course and an inside brick course were laid up and filled in between with brick batts. Sand for brick mortar was hauled from other excavations, and no attempt was made to wash it.

2.9 Miscellaneous Concrete Failures

Some concrete failures do not easily lend themselves to categorizing. They may be attributed to any one of several causes. Three such are described below.

Johnson Electric Service Co.—Milwaukee, Wis.

On March 28, 1903, the forms under the freshly poured building[31] for the Johnson Electric Service Co. in Milwaukee, Wis., gave way and an area 18 × 55 ft fell, taking all the lower floors with it. No damage resulted in any other parts of the structure. Later investigation, however, indicated that the failure may have been the result of faulty brickwork in that as many as fifteen courses of brick were counted without any bond between the inside 8 in. and the outside 18 in.

Factory building—Hersfeld, Germany

Despite the fact that the Europeans are not so much in a hurry as we Americans are, they also experience concrete failures. This one, reported in February 1923, is interesting because of the way in which it was brought about.

The frame of a two-story reinforced-concrete factory[32] in Hersfeld, Germany, had been completed in February 1923 and all forms left in place for proper curing of the concrete. While the concrete was setting, and with no workmen on the job, a steel erector working on an adjacent building attached his tackle to one of the concrete column forms to lift some 70-ft trusses. The lateral pull on the tackle put enough pressure on the forms and, through the forms, on the green concrete, to fracture the concrete column. When the forms were removed, a large diagonal crack showed in the column near the base. An hour later the structure collapsed, killing five workmen.

Atlas Steel Co.—Welland, Ont.

During World War II, any accident that could not be readily explained would raise the charge of sabotage. Often, the accident probably could have been due to causes connected with the war effort although not sabotage—perhaps excessive speed, or the necessity to use incapable or inexperienced help because of the manpower shortage.

In 1942, at the Atlas Steel Co.[33] at Welland, Ont., concrete was being placed in a 6-in. slab, 20 × 60 ft. The formwork, supported on jack shores, gave way, killing 4 men and injuring 31 others. The cause was unknown, but the press stated that it appeared some one had tampered with the jacks. Was it sabotage, or inefficiency?

3

Steel Failures

A steel structure is composed of shop-fabricated members made up of a material whose strength can be controlled within fairly accurate limits. Steel failures, therefore, are seldom caused by any fault of the material as such, but primarily by the actual erection or guying of the structure. The high proportion of this type of failure described here will emphasize this fact. Several of these bracing failures received an assist from the designer.

Two other closely related types of failure exist: (1) the failure of a steel frame not yet properly braced, when hit by winds of gale velocity; and (2) the failure of the frame through an accident; for example, a truss in place is bumped by a truss being erected. The author knows of no steel failure in a completed building that could actually be classed as a failure of the steel.

3.1 Specific Lateral Weakness

The failures under this heading are described separately because there was a specific weakness in design that induced a lateral weakness. Most of the structures were sound except for one error in design or judgment, which may have been an oversight in providing for proper lateral bracing, or an error in some specific detail. In any case, the end result was the failure of the building.

First Christian Church—Long Beach, Calif.

The top fibers of a simple beam are stressed in compression (Sec. 1.3). A truss, however, has no web to carry a decreasing compression down as far as the neutral axis to the bottom fibers. All of the compression is concentrated in the top chord member of a truss, and all of the tension in the bottom chord. For all intents and purposes, therefore, the compression flange of the truss becomes a column and is subject to the same limitations of slenderness ratio as a column (Sec. 1.4). Therefore, to carry a stress efficiently, the top chord must be supported laterally to prevent it from buckling sidewise. The lack of this lateral stiffness for the top chord of the truss, which was provided in the original building design, caused the failure of the

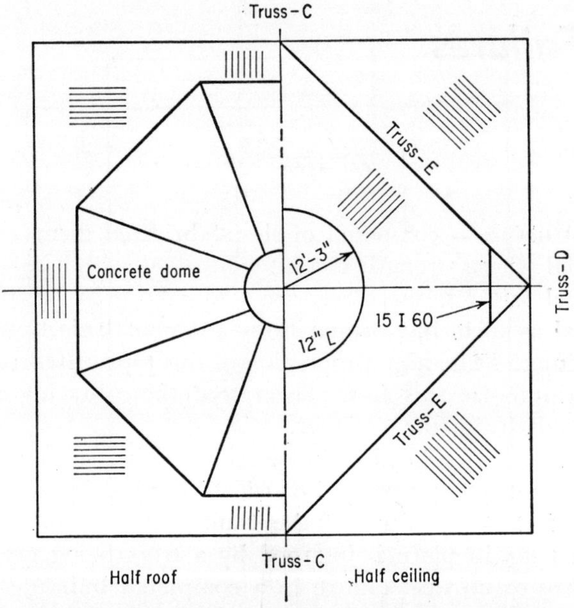

FIG. 3.1. Plan of dome for First Christian Church

First Christian Church [34] at Long Beach, Calif., December 16, 1920. The failure occurred during the lunch hour, and the casualties were limited to a few minor injuries.

In general, the structure of the dome of the building, the portion that failed, consisted of a segmental concrete dome, made up of eight segments, 48 ft in diameter, rising to a height of 76 ft above the floor (Fig. 3.1). The shell of the dome was 3 in. thick, reinforced with steel fabric and rods, and weighing about 200 tons. The dome rested on an 8-in.-thick vertical concrete wall, which was carried by a framework of steel beams and trusses, shown in Fig. 3.2 as four E-trusses, and four 15-in. I-beams. The E-trusses, in turn, were carried by four reinforced concrete columns and formed a 62-ft square.

The top chords of the E-trusses were braced against the top chord of the C- and D-trusses by 6-in. H-sections, and the top chord of the C-trusses were braced on one side to the outside wall with 20-in. steel beams; and on the opposite side by 24-in. beams, both of which were framed, not to the top chord, but to a vertical member of the truss at a point about 3-ft above the bottom chord, so that the top chord lacked any bracing against lateral buckling.

The D-trusses, in turn, were braced to the outside wall with 15-in. beams, which had originally been shown to frame from the top

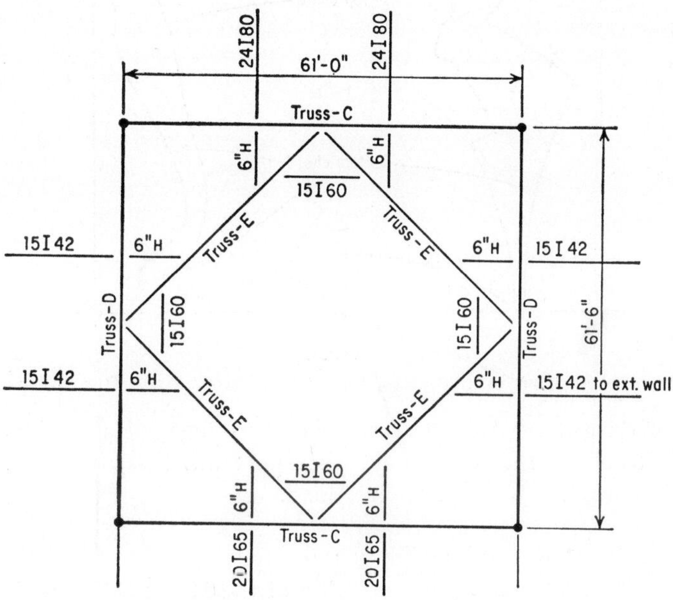

FIG. 3.2. Position of framing of First Christian Church before collapse

chord to the wall, but which had been dropped to an elevation several feet lower instead and hung by a gusset plate. The application of the load in this manner did two things—it removed the main lateral bracing of the top chord of truss-D, and it applied the load in such a manner as to induce a tipping action of the truss; the actual failure was precipitated by the tipping and buckling of the D-truss.

When the failure occurred, the building was almost completed. All the concrete was in place and the forms removed, but 60 tons of outer stone covering had not been placed on the dome. The collapse was strictly a steel failure: the concrete columns supporting the trusses remained in place, and the concrete of the dome was in remarkably good condition after the failure. The position of the trusses after collapse is shown in Fig. 3.3.

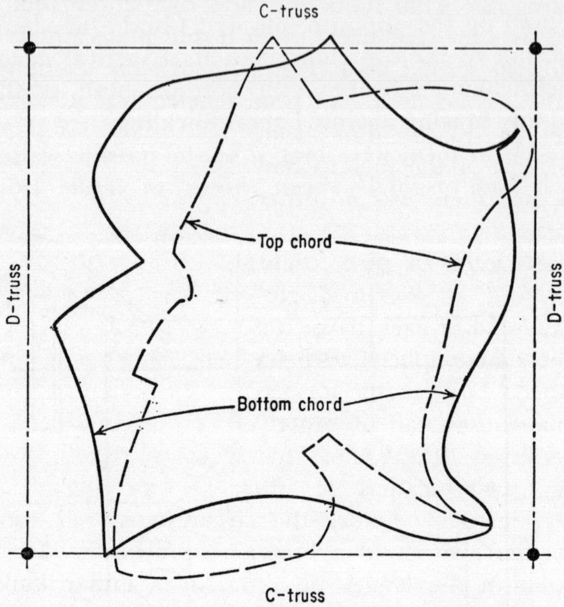

FIG. 3.3. Position of trusses of First Christian Church after collapse

It later developed that the contractor had engaged an independent engineer to check the design and his report was that, in his opinion, the design was weak against wind load.

Hyatt Buick Building—Kansas City, Mo.

The steel failure during the erection of the Hyatt Buick Building [35] at Kansas City, Mo., January 29, 1917 occurred during the noon hour,

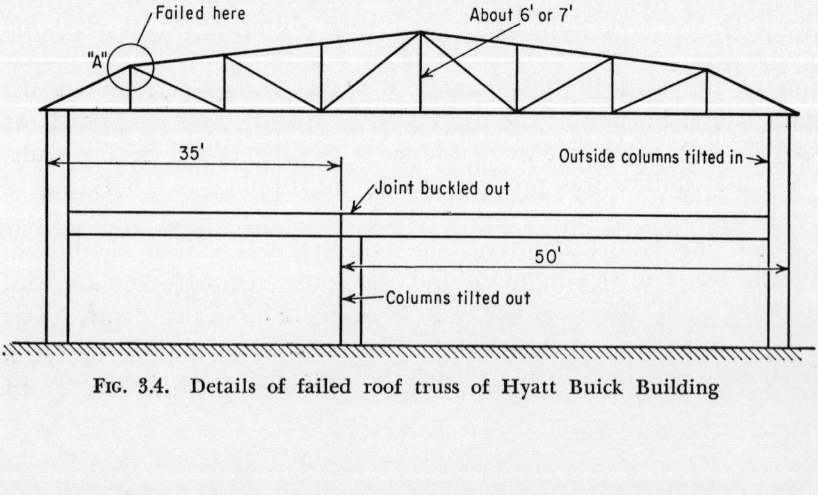

FIG. 3.4. Details of failed roof truss of Hyatt Buick Building

when no workmen were on the job, which probably was the saving feature in avoiding any injuries or deaths.

The building, 85 × 130 ft in area and two stories high, had bowstring steel trusses for roof spans of 85 ft and plate girders of 35- and 50-ft spans for the second floor (Fig. 3.4). All the columns and plate girders for the second floor had been erected but none of the floor beams were in place. The roof trusses and a few of the purlins had been placed, but there was no other lateral bracing.

The trusses were of very light construction, and at the first gusset in from the end of each truss, top chord members were spliced to form the bowstring. This point formed a vertical plane that had absolutely no lateral stiffness other than that furnished by the thickness of the gusset plate (Fig. 3.5). Under the circumstances, it was not even necessary for the whisper of a breeze to precipitate the failure. A small variation in the length of a purlin—a minor kink—anything that would produce a slight eccentricity in any one of the trusses was sufficient to start the collapse.

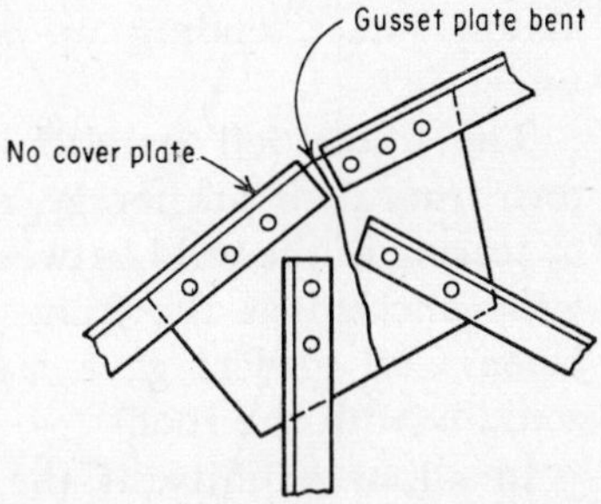

FIG. 3.5. Detail "A" in Fig. 3.4

Sem Industrial Building—Westbury, N.Y.

The necessity for recognizing that sidesway must be provided for was demonstrated in the failure of the Sem Industrial Building[36] in Westbury Industrial Park at Westbury, N.Y., in May 1959. In this one-story building, about 15 ft high and covering an area of 175 × 143 ft, the girders were run over the top of the columns with knee braces from the girders to the columns. The beams ran continuously over the girders but knee braces were not placed parallel to the beams.

All steel had been erected and the roof had been completed but no masonry was in place when the entire structure tipped over in the direction of the beam framing, killing one workman and injuring several others. The structural design was the same as that of 25 other buildings in the Industrial Park, and had been approved by the town building department.

The columns were 6-in. H-sections, spaced alternately 25 ft and 33 ft in the direction of the girders and 25 ft in the direction of the beams. The columns had plates welded top and bottom, anchored to the foundation by means of two ¾-in. anchor bolts, and at the top, the girders ran continuously over the columns, extending several feet as cantilevers supporting shorter simple span girders.

The beams, spaced 8 ft on centers, extended over and were bolted to the top flange of the girders by means of ½-in. bolts. The steel framing had no permanent bracing, but cable guys had been used during erection. The roof was a 2-in. poured gypsum, supported by bulb tees over the beams. When the building fell, the roof remained almost intact, ending up at ground level and trapping workmen under it.

The frame fell parallel with the beams and the columns were torn from their anchorage, either by stripping from the anchor bolts or tearing of the weld between the column and the base plate. Girder webs buckled as the frame tilted over, and the bolts between the beams and girders gave way, permitting the beams to move horizontally with the roof.

In all probability, if the structure had stayed up until the side walls had been built, there would have been no further damage. If the other buildings in the park are built from the same design, they are undoubtedly stiffened by the masonry.

When the building was rebuilt, several changes were made in the design. Girders were run from column to column and frame in instead of running over the top and knee braces were placed in both directions. Moreover, the local building department, as a result of this failure, wrote in a number of new code requirements including: (1) construction supervision and certificate of approval to be supplied by a registered architect or licensed professional engineer; (2) all buildings to be designed for a 20-lb wind load in all directions; and (3) all main compression members to be designed to keep the ratio of slenderness under 120, and in case of any column unbraced in either direction, the design length of the column used in determining the ratio of slenderness was to be double the actual length.

Pacific Nut Oil Co.—Los Angeles, Calif.

The failure of the steel frame of a building [37] under construction for the Pacific Nut Oil Co. at Los Angeles, Calif., June 23, 1937, is a repetition of cases already described, but this repetition will emphasize the fact that the main reason for failure of a steel framework is lack of bracing.

The 1-story warehouse building, with an area of 150 × 300 ft had as its main roof supports a series of eighteen steel trusses of 100-ft span. No diagonal braces had been provided, and the trusses were held in place by guys to the north only—no counterbracing was provided toward the south. All the steel was up and the erectors were in the process of plumbing up (check for true vertical alignment) the steel. The cables to the north were released, probably to plumb the trusses, and the entire roof structure came down as a unit. Two men who were on the trusses rode them down and were killed in the collapse.

Investigation disclosed the facts that the framework was as much as 5 in. out of line and that, in places, only a single erection bolt was in place where holes had been provided for three.

Brooklyn Theater—Brooklyn, N.Y.

A typical instance of a structure that, if properly braced, would have been perfectly safe, was the Brooklyn Theater,[38] which failed November 29, 1921, causing the death of seven men. The building, about 80 × 160 ft in plan of the main portion, was designed for

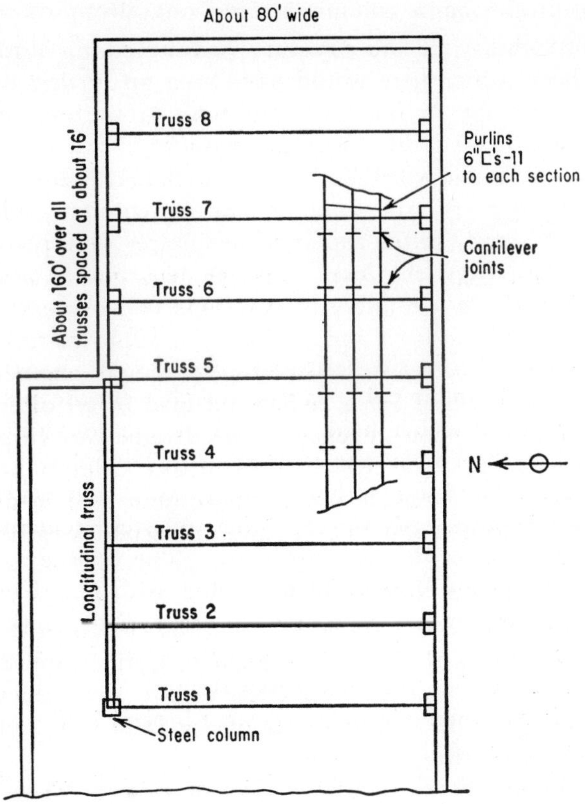

FIG. 3.6. Approximate plan of Brooklyn Theater before collapse

a seating capacity of about 1,800. The transverse trusses, 1 through 8 in Fig. 3.6, were spaced about 16 ft on centers, and over the top of the trusses were 6-in. channel purlins, cantilevered out about 20 in.

as shown to pick up short purlins on a simple span. Over the trusses was the roof of reinforced clay tile. Four of the eight trusses were carried on brick piers built integral with the wall. Three of the others (Fig. 3.7) were carried on top of a longitudinal truss (Fig. 3.8), truss 1 being supported on a double strut to the top of the column which carried the longitudinal truss.

Apparently the only faulty design, aside from the lack of bracing, lay in the fact that this main column was not anchored to the concrete footing. Although holes were provided in the base plates of the transverse trusses, they were not anchored to the pilasters on which they were carried. There was no diagonal bracing whatsoever between the trusses and there was one line of strut bracing near the center of the lower chord.

Investigation showed that the collapse had been precipitated by the base of the main column kicking out, dropping all of the

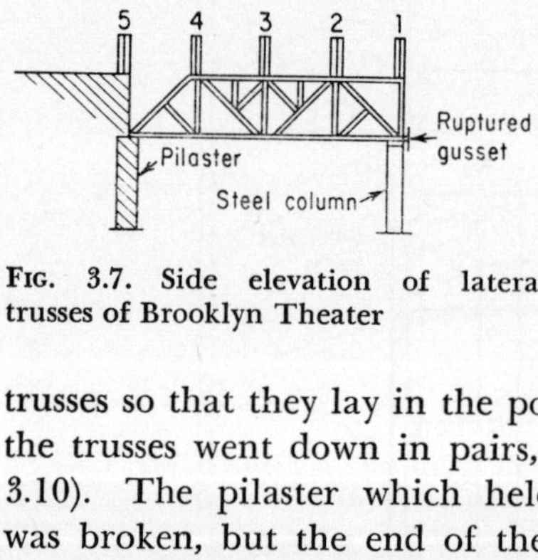

Fig. 3.7. Side elevation of lateral trusses of Brooklyn Theater

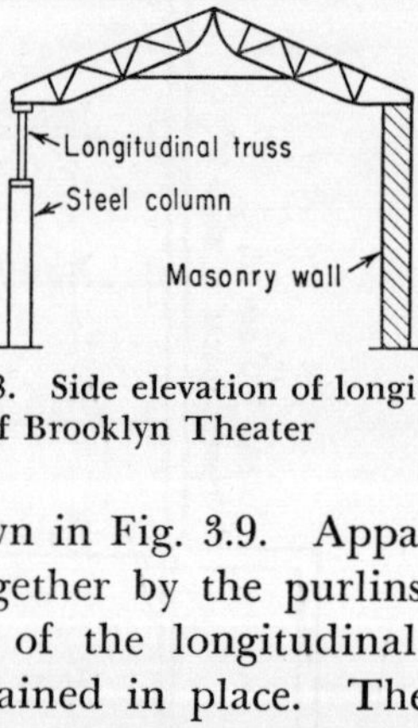

Fig. 3.8. Side elevation of longitudinal truss of Brooklyn Theater

trusses so that they lay in the position shown in Fig. 3.9. Apparently the trusses went down in pairs, as tied together by the purlins (Fig. 3.10). The pilaster which held the end of the longitudinal truss was broken, but the end of the truss remained in place. The wall of the offset fell, and one other pilaster, as shown on the plan, (Fig. 3.9) was broken. All of the other pilasters remained intact and the trusses slid off the bearing plates. The double angle strut, which supported truss 1, was bent double with the back of the angles against each other. The trusses scraped the inside of the pilaster while falling in only three places in the building. In all others, the end of the truss fell completely clear of the wall. In general, very few of the main members were broken in the fall, except in the web members of the trusses.

It is difficult to determine what triggered the failure. There was some indication that the failure of the pilaster at the north end of column 5 may have caused a slight jolt which kicked the column out at the far end of the longitudinal truss, or it may have been that some extraneous cause brought about a vibration which jolted the column and caused the collapse.

This was another instance of an owner buying a set of plans without any architectural or engineering supervision and then proceeding to act as his own contractor. As frequently happens after such an accident, an immediate hue and cry arose to require professional inspection on all building construction of this type, but, as usually happens, the outcry subsided soon after.

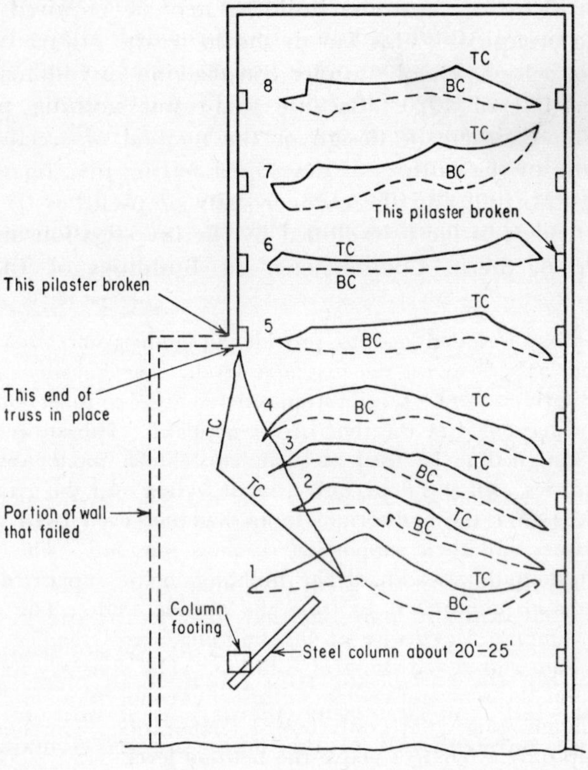

FIG. 3.9. Approximate plan of Brooklyn Theater after collapse

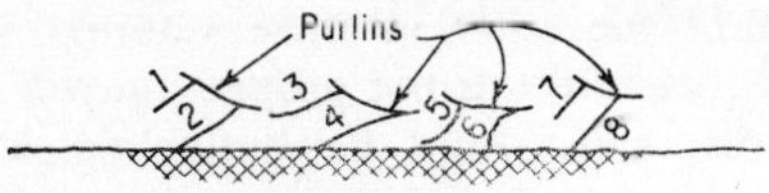

FIG. 3.10. Cross-section showing position of trusses after collapse

3.2 Normally Safe Bracing that Failed

Sometimes bracing that would normally have been considered perfectly safe is subjected to treatment from some outside source for which it was never intended. The results of these failures are usually reflected in the insurance rates of the contractor, but it is hard to point a finger at the person responsible. These cases are usually written up in the newspapers and listed as construction accidents.

High school—Indianapolis, Ind.

The statement has been made or demonstrated in other sections of this book that the stresses encountered during erection may be far greater than any that the building may be required to sustain after being occupied. This fact is shown in the failure of the steel structure of a high school [39] under construction in Indianapolis, Ind. during the fall of 1920. Because there was nothing particularly unusual about either the design or the method of erection, a complete picture of the failure—as developed during investigation by the building authorities and the coroner—may be useful.

The statement of facts developed by the investigation and released by Walter B. Stern, Commissioner of Buildings of Indianapolis, follows:

The building was designed to provide an auditorium 90 × 106 ft and a gymnasium 54 × 90 ft on the first-floor level. The balconies at the sides of the auditorium were of reinforced-concrete tile construction and were in place and hardened at the time of the accident. The entire third floor, which was designed to be used as a students' lunch room, was suspended from roof trusses. Of this floor, only a small section over the gymnasium was poured, while over the auditorium forms had not even been placed. All the roof trusses and their supporting columns were up. The trusses over the auditorium ran crosswise of the building, being supported on trusses on steel columns about 9 ft in from the outside walls. The trusses over the gymnasium ran lengthwise of the building—across the 54 ft width of the gymnasium—and rested on steel columns. The trusses were completely riveted up but the roof and ceiling purlins, the connections to the columns, and the column splices were only bolted temporarily. The side walls had been laid up only a few feet above the balcony level.

Witnesses at the coroner's investigation, conducted with the assistance of representatives of the city's building, police, and fire departments, testified that more than half of the holes in these various connections had bolts in them but the steel structure was not properly guyed to provide the necessary bracing until the walls could be built high enough to bear the structure.

A wooden stiff-leg derrick with a 32-ft mast and with a 12 × 12-in. wooden boom 81 ft long had been placed on the top chord of one of the trusses five or six days previous to the accident. [Fig. 3.11]. The derrick was rigged and in service on the morning of the accident. The brake lining on the boom hoist drum of the hoisting engine was practically worn out, and it is probable that the boom was handled in a jerky manner when being lowered.

It was determined at the investigation that the cause of the accident was the improper rigging and operation of the derrick before the steel structure was adequately braced and riveted. The roof trusses were designed to support a load in the finished building many times the weight of the derrick. Testimony given at the investigation supports the view that an

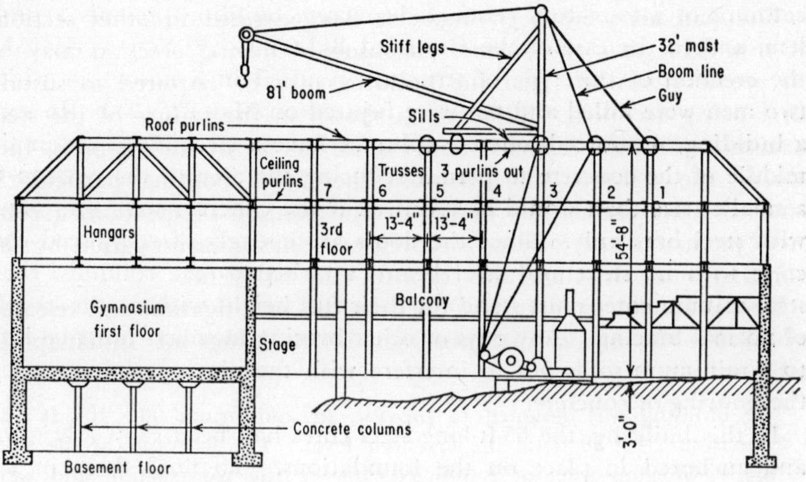

FIG. 3.11. Longitudinal section of Indianapolis High School before collapse

unusual thrust, caused by the operation of the derrick, sheared the connection between a roof truss and its column—causing the truss to start down at one end. The pull on the purlins then brought about the collapse of the remainder [Fig. 3.12].

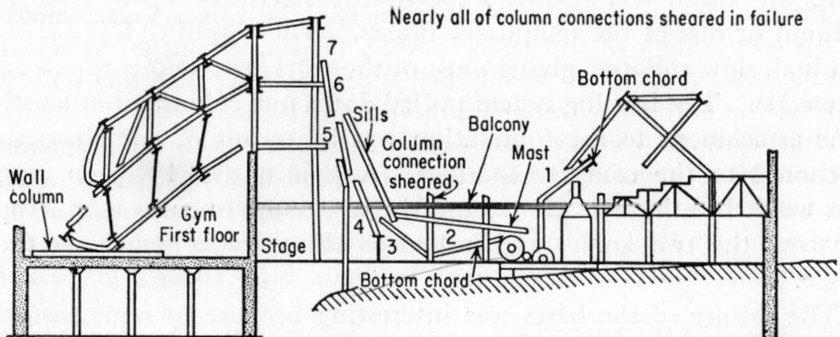

FIG. 3.12. Longitudinal section of Indianapolis High School after collapse

The collapse is an instance of a type of failure that could happen to any of us. The unfortunate fact is that, regardless of the innocence of the designer or anyone else, regardless of any report clearing the parties concerned of any negligence, there may be a certain stigma in the minds of the general public that is bound to react against the designers of the structure.

Gottfried Bakery—New York, N.Y.

The erection of the steel cores for composite concrete and steel columns of a four-story reinforced-concrete building presents a problem, and in the case of the Gottfried Bakery Co.,[40] New York, N.Y., the erection of this type of structure resulted in a failure in which two men were killed and ten were injured on May 22, 1930. In such a building, reinforced-concrete columns have a steel H-section in the middle of the concrete to permit reducing the area of the column to a smaller size than would be required if the concrete were reinforced with steel bars only. Since the floors are entirely of reinforced concrete with no structural steel beams attached to the columns, these steel column cores must stand for their full height without the benefit of normal bracing. Any guys or other bracing members must be kept to a minimum so as not to interfere with the placing of forming or the pouring of concrete.

In this building, the 65-ft-long steel cores had been shop fabricated and anchored in place on the foundations. To hold them in the proper position for concreting, temporary bracing beams were attached to the columns, 6½ ft above the third-floor level in such a way as to permit the pouring of the third and fourth floors. These beams were 6-in. I-beams and channels in both directions. In addition, diagonal guy cables were used in some places.

At the time of the failure, all of the steel cores had been erected, and the last derrick was being dismantled. In the dismantling operation, the boom fell against some of the steelwork—against either a column or one of the temporary braces. The resulting failure was a gradual, slow collapse, giving most of the workmen ample opportunity to escape. The bracing system pulled down one column after another. The attachment to the foundation was by means of steel base slabs anchored to the concrete footings. A pair of $6 \times 4 \times 3/8$-in. angles was welded to the base plates, and the steel column cores were slipped between the two angles and bolted to them, using two bolts to an angle.

The failure of the bases was interesting because in some instances the welds broke, in others the bolts sheared off, and in a few instances, the entire column footing was uprooted and turned over with the column.

Bayside High School—New York, N.Y.

Seldom is a steel framework actually knocked down, but that is literally what happened at Bayside High School [41] in New York, N.Y., December 13, 1934.

A bay of steel framing, 13 ft wide by 75 ft long, ranging from one to four stories in height, was not connected to the rest of the building by any framing, and, therefore, relied entirely on the anchor bolts and guying for stability. A crawler crane was lifting a bundle of steel beams, which, in swinging around, hit the framework in its narrow dimension, knocking over 30 tons of steel that had been erected and killing two erectors who were on top of the structure. The rest of the building was undamaged.

School auditorium—Everett, Wash.

Probably more failures of the type described here occur that do not get into the newspapers than those we read about. An accident during construction operations in which nobody is injured does not rate much publicity.

On May 19, 1939, the last one of five 113-ft long trusses over an auditorium-gymnasium of a school [42] in Everett, Wash., was being swung into place when it struck a truss already in place but not yet braced. The truss collapsed slowly carrying with it the two adjacent spans. Observers said that the collapse took a full three minutes, giving workmen ample time to escape.

Stock Show Building—Ottawa, Ont.

A steel erection failure, similar to so many others, occurred April 23, 1906, when the trusses of the new Stock Show Building,[43] being erected at Landsdowne Park, Ottawa, Ont., fell, injuring sixteen workmen.

Six trusses had been erected and fastened together by purlins, three between each pair of trusses. Heavy diagonal bracing in the plane of the outside columns was called for but had not yet arrived, so the steel was being erected without it. At the time of the accident, planks for a scaffold were being raised by a pulley attached to one of the trusses. Apparently the pulley fell, allowing the scaffold lumber to fall on one of the guys, which pulled over the trusses one after the other.

Strangely, the column that was anchored to the concrete foundations bent over without breaking any of the anchor bolts or the concrete. This was the third collapse of the same building during its construction period of slightly more than two years. The two previous failures had been attributed to snow load and to faulty construction. The first occurred January 25, 1904, injuring four men. The second failure occurred January 19, 1905, when the building was almost completed. This time eight trusses fell, leaving five others in place. The frame was then completely dismantled and a new contract awarded.

Beaver Area High School—Beaver, Pa.

Even a well-braced steel structure being erected by a capable erector can fail, as indicated by the collapse of four of a series of 204-ft span steel arches of the Beaver Area Senior High School [44] at Beaver, Pa., in July 1959. The arches constituted the main roof members of a 110 × 205-ft gymnasium.

The two-hinged steel arches were spaced 27 ft on centers and had a rise of 37 ft 3 in. The structure was designed to have five arches, with the center arch a heavier section to support a folding partition. In order to carry this folding partition, a beam spanned opposite sides of this center truss. The purlins between the four arches that were in place had been erected and knee braces were being placed from the 14-in. wide flange purlins and the beam which carried the folding doors, to prevent lateral movement when the door was operated. The arches were well guyed in both directions, although the completed section was considered to be thoroughly rigid as it stood, without dependence on the guys. Most of the knee braces were in place, but to install one of these knees, it became necessary to exert an external force, apparently large enough to cause one of the cable clamps to slip on one of the guys, thus precipitating the collapse. Five steel erectors who were in the structure when it failed rode the steel down and were injured in the failure. No one was killed.

3.3 Overload on Steel Frame

Once in a while a report comes through about a steel structure that failed because of overload of construction materials. Probably for every failure of this kind that gets into the newspapers or magazines ten others occur—frequently minor failures.

In addition, probably 50 or 100 to 1 are greatly overloaded, yet never quite arrive at the point of failure. Probably more than half of the steel structures that are built are, at some time and in some part of the building, overloaded. Two of these overload failures have been included in this section.

Theater under construction—South St. Paul, Minn.

Crowded storage conditions frequently lead a contractor to take chances in construction. On the night of August 4, 1914, a theater [45] under construction in South St. Paul, Minn., collapsed under the weight of 25 tons of steel piled on an upper area. It had been hoisted during the afternoon. Since the failure took place after working hours, there were no casualties.

O'Brien, Potter, and Stafford Building—Buffalo, N.Y.

An attempt to mix and pour concrete by gravity from the roof of an 8-story steel frame and bar joist office building would probably have resulted in a saving to the contractor, if it had not been for lack of proper supervision. In this case one interior bay about 20 ft square of the O'Brien, Potter, and Stafford Building of Buffalo, N.Y., failed from the roof through to the basement, April 14, 1930. Three men were killed and several others injured.

The building was of steel skeleton construction with steel bar joists and concrete floor slab, properly designed and accepted by the city building department. Transit-mix concrete was not generally used then and the building occupied a busy downtown corner, so space for mixers, storage of material, and the like was at a premium. Under the circumstances, it was not difficult to sell the contractor on a mixing system that hoisted the materials to the roof, dropped them to the eighth floor for mixing, and then, by means of chutes, placed the concrete where it was needed on the floors below. The system depended upon uniform delivery of materials to the roof and uniform use of the concrete so that materials would not be stored on the roof framing. The bar joists of the roof were designed for a total live and dead load of 100 psf; when planked over, this allowed for only a shallow pile of materials over part of the bay.

The contractor in charge of the job had seen to it that the mixer was operating smoothly and that the cement and gravel were arriving at proper intervals and were being hoisted to the roof in barrows. Then he left the job briefly for an appointment. Almost as soon as he had left, something went wrong with the mixing equipment and it was stopped for repairs, but nobody thought about stopping the delivery of cement and gravel and it began to pile up on the plank platform on the roof. When it had accumulated to a load of approximately 350 psf—three-and-one-half times the design load—the bar joists could take no more. They gave way, carrying the mixer through from the eighth floor and all the materials from lower floors, including a hoisting machine on the first floor, to the basement.

In the reconstruction of the building, no change was made in the design, but conventional methods of mixing and placing concrete were used.

3.4 Structural Frame Failure Caused by Wind

Despite all normal precautions, during certain stages of construction a building is vulnerable to the forces of nature. If a steel frame is up, but held together only by means of erection bolts and guys, it does not require a wind of hurricane force to loosen the guys and wreck the frame.

Many of these failures occur during the hours when no workmen are on the job, and, therefore, this type of failure does not cause as many deaths as many other types, and so receives less publicity.

Garage building—Chicago, Ill.

On November 14, 1927, a partly erected steel frame for a three-story garage building [46] in Chicago, Ill., was blown over, injuring ten people who had taken refuge in the contractor's shanty, and two others in a passing automobile that was crushed by the falling framework.

The erection and temporary bolting were finished Saturday, November 12, with all holes filled and all bolts tight. On Monday morning the structure was being plumbed preparatory to riveting when a heavy wind and rainstorm came up. The second-floor level had plate girders 40 in. deep and about 55 ft long, which provided a large area exposed to the wind. No diagonal bracing or guys had been placed to restrain the structure, and it was entirely dependent on the anchor bolts and the rigidity of the bolted connections for wind resistance. The concrete was still rather green and, at every column base, either the anchor bolts broke off, the threads were stripped, or the footings at the ground floor level yielded to permit the failure.

H. D. Pierce Building—Oswego, N.Y.

A little knowledge is a dangerous thing, and the supervision of steel and concrete construction is no job for an amateur.

The roof of a two-story building,[47] 66 × 100 ft in plan, under construction for H. D. Pierce at Oswego, N.Y., fell October 22, 1915, when the roof concrete was being poured. The construction of steel frame, hollow tile walls, and concrete floors and roof had steel roof trusses 16 ft 5 in. apart, and plans had been approved by the building department. All the members in themselves were sufficiently strong, and all were properly guyed. Shortly before the failure, the owner took over supervision of the work from the architect, and subsequently permitted removal of one of the guy cables to allow better access.

A 20- to 25-mph wind was blowing, and in all probability the eccentricity of loading, while concrete was being placed, together with the vibration of placing, contributed to the failure. The trusses, lacking bracing, turned over on their sides, twisting the tops of the columns with them. The collapse killed one man. It is interesting to note that the second floor sustained the shock without any failure. The only damage to the walls was at the column locations where they had been pulled out of the masonry.

Lankenau Hospital—Philadelphia, Pa.

On Sunday afternoon, December 17, 1951, 75 tons of steel, constituting the framework for one wing of a new hospital,[48] collapsed during a windstorm. The framework, situated on a hill in the outskirts of Philadelphia, Pa., had been bolted and was ready for riveting. Riveting crews reported for work on Saturday morning but were unable to work because of rain. The Weather Bureau reported gusts of 40-mph velocity on Sunday afternoon when the structure fell.

Apparently this gusty condition set up vibration in the structure, which loosened up the bolts, allowing one or more of the girders to slip, thus precipitating the collapse. Few of the members were distorted or otherwise damaged. No one was injured.

3.5 Eccentric Load on Columns During Erection

Although neither of the two items listed resulted in a major failure, they are so much alike that a simple lesson may be learned from them. In both instances, serious injury resulted to the steel erector involved.

Whitmer Jackson Co. Storage Building—Buffalo, N.Y.

On June 6, 1950, while steel was being erected on the Whitmer Jackson Co. Storage Building in Buffalo, N.Y., a steel erector was injured when a column tipped over. The circumstances of the failure brought about a lawsuit between the compensation insurance carrier for the steel company and the general contractor—represented, of course, by his liability insurance company.

The plans, as filed with the city building department, showed the steel columns approximately 12 ft high anchored to concrete footings which were 2 ft square and 1 ft 10 in. deep from the underside of the 5-in. floor slab.

When one of the erectors climbed up the column to make the connection of a beam that was swung into place, the eccentricity of his weight, when he was about two-thirds up its height, tipped the column. The anchor bolts did not break, but the footing tipped with the column. On measurement, the pier was found to be 1 ft 6 in. square, instead of the 2 ft called for, and 10 in. deep at the center with a rounded bottom, so that at the sides the depth was as little as 6 in. On the other hand, the steel erector did not guy the column, nor in any way attempt to provide for eccentric erection stresses, depending entirely on the mass of footing to resist tipping the column. In view of the negligence of both parties, the case was settled out of court.

A suburban high school column in western New York

The author was called in by an insurance company lawyer on this investigation, mostly to advise on custom and usage.

The general contractor had been required by the specifications to place the anchor bolts for steel columns in accordance with the locations and elevations shown on the shop drawings. The ordinary method of placing the anchor bolts is to nail a 2 × 4 wood piece flatwise across the top of the form for the concrete footing. Holes are drilled accurately into the wood to receive the anchor bolts, which are held at the correct elevation by means of nuts threaded down on the top of the anchor bolts (Fig. 3.13).

For some unknown reason, in this instance, the general contractor welded the turned-over end of the anchor bolt to the mat of steel reinforcement. When the concrete was poured, the mat was depressed by the weight of the concrete and the men working the concrete around the bars, pulling the anchor bolts down with the mat. When the time came to place the steel column and base plate, there was scarcely enough of the anchor bolt left above the base plate and clip angle to catch more than one or two threads of the nut.

The steel erector suggested that the anchor bolts be extended by means of a threaded rod, attached by a sleeve nut, but the general contractor preferred to weld on the threaded extensions instead. For the first few columns, the general contractor employed a local welder to do the job. When the next load of steel columns arrived, the welder was not available and, to expedite the erection, one of the steel erector's employees—who had some welding experience—undertook to do the welding, using the local welder's equipment.

The stubs were welded on and the columns dropped into place over the anchor bolts and the steel base plate, which had previously been grouted in place. The nuts were turned down on the anchor bolts, thus firmly anchoring the columns in place.

When the load of steel beams arrived, the first beam was lifted into place, and the two erectors climbed up the two columns—the last one previously erected and the first of the new lot—in order to attach the beam to the column. When the erector got to the top of the new column and started to reach out for the beam, the eccentricity of his weight broke the weld of the anchor bolt stub, and the column fell, carrying the erector down with it. One of his legs was sheared off between the edge of the concrete wall and the column in the fall.

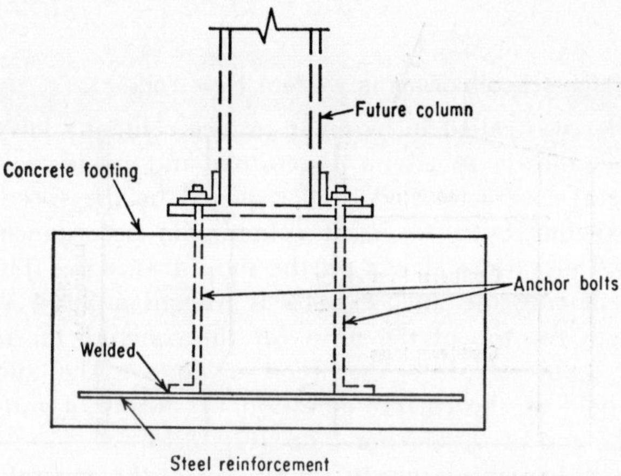

FIG. 3.13. Cross-section of column and footing showing improper procedure in anchorage

The only question was one of responsibility. In accordance with the specifications, it was the responsibility of the general contractor to prepare the anchor bolts to tie the columns safely and accurately. However, when the employee of the steel erector undertook to do the welding, he transferred this responsibility to his employer.

Orpheum Theater—New York, N.Y.

In more than one famous collapse listed in this book, the experts have disagreed as to the real cause of failure. In the failure of the Orpheum Theater [49] in New York, N.Y., January 22, 1913, the design had a number of flaws, no one of which, by itself, would have caused the failure. In all probability, buildings are standing today in which a slight variation of all of the listed discrepancies exist. The failure occurred 20 minutes after quitting time and, therefore, nobody was injured—the only loss was to physical property. Probably as a result of this failure, many other buildings have been better built.

The building was a large-capacity theater, designed by the foremost theater architect of his time, Thomas W. Lamb. That the fallacies in the structural design were not evident is proven by the fact that the building department examined the plans and approved them. Moreover, the materials and workmanship were good. Yet, when the building was in the stage of plastering operations, the greater part of the roof and balcony collapsed.

Six investigations were begun and carried on independently—one on behalf of the steel contractor, four separate ones on behalf of the general contractor, and one by the bureau of buildings. Although

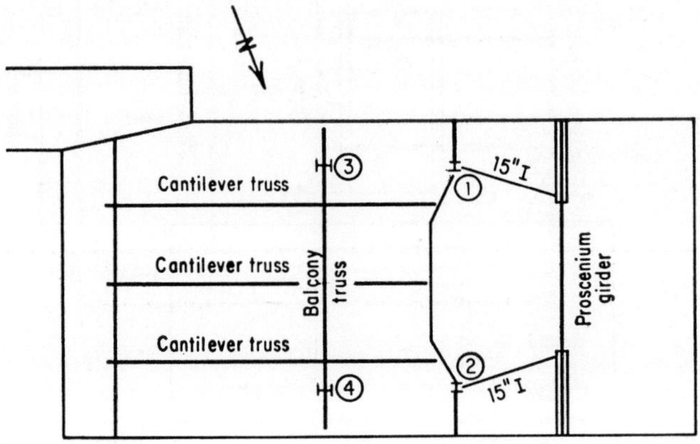

FIG. 3.14. Balcony framing plan of Orpheum Theater

each investigator was willing to admit that the other causes suggested may have contributed to the failure, no one of them accepted any of the others completely, probably a healthy sign.

The general layout of the steel structure is indicated in its main parts in Fig. 3.14–3.17. The roof was of concrete, carried by beams into a system of trusses—a long wall-bearing truss (truss B) and another truss (truss A) toward the proscenium end carried on columns 1 and 2, with trusses at right angles between the two major trusses. The proscenium wall also carried roof beams. The portion which fell was all of that section from the proscenium back to truss B, and the balcony below this area. The balcony was of cantilever trusses carried through a main truss (Fig. 3.16), supported on short columns 3 and 4. The fascia girder of the balcony curved from the ends of the cantilever trusses to columns 1 and 2, which carried the roof truss.

The main points of failure uncovered were: (1) in the main balcony truss, the bottom chord was broken at the center field splice—the top chord at this point bent and was partly broken, and several web members were bent or broken; (2) the truss remained attached to columns 3 and 4 and both columns either buckled or tore at

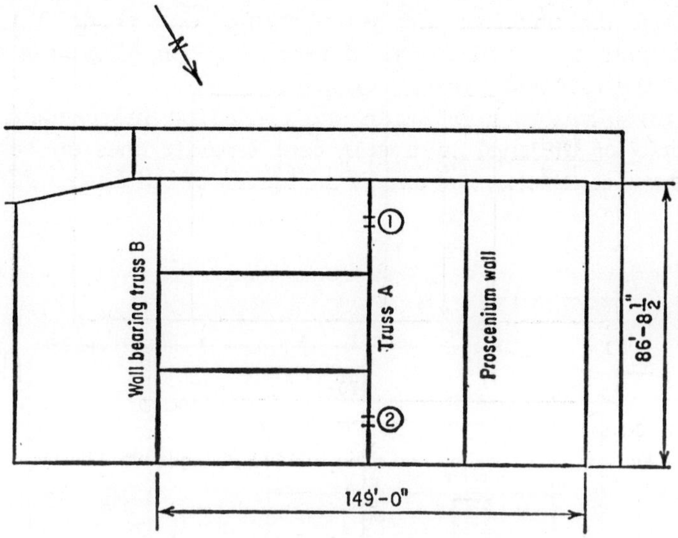

FIG. 3.15. Roof truss plan of Orpheum Theater

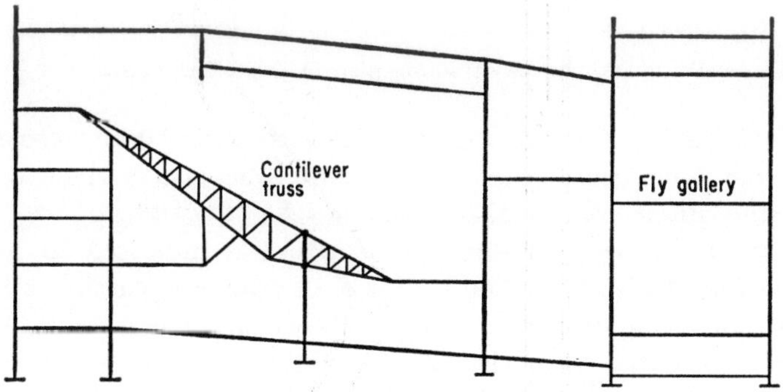

FIG. 3.16. Longitudinal section of Orpheum Theater

the bottom chord connection; (3) the roof columns 1 and 2, 60 ft long, were braced in one direction only at two points, where the balcony fascia beam framed in, in one direction, and where the 15-in. I-bracing beam was framed to the proscenium girder. The columns buckled at both bracing points and at the bottom of truss A, but did not become detached from the truss (Fig. 3.18).

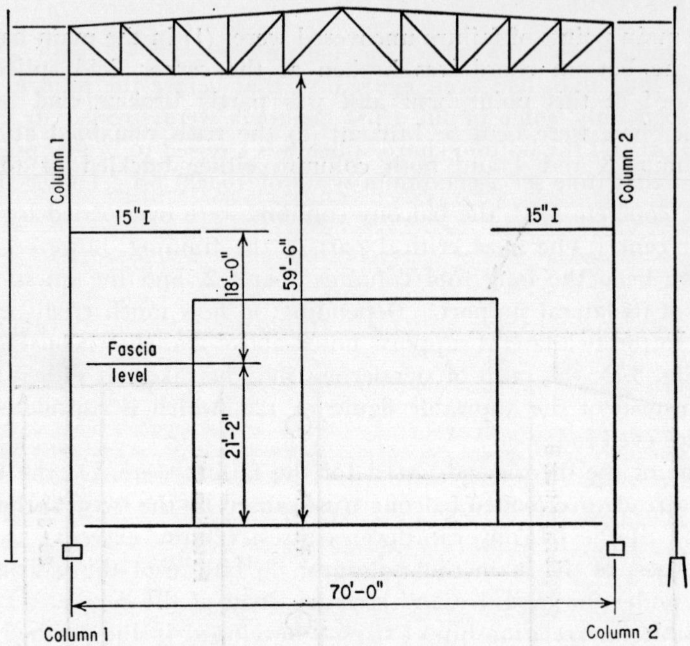

Fig. 3.17. Proscenium and main columns of Orpheum Theater

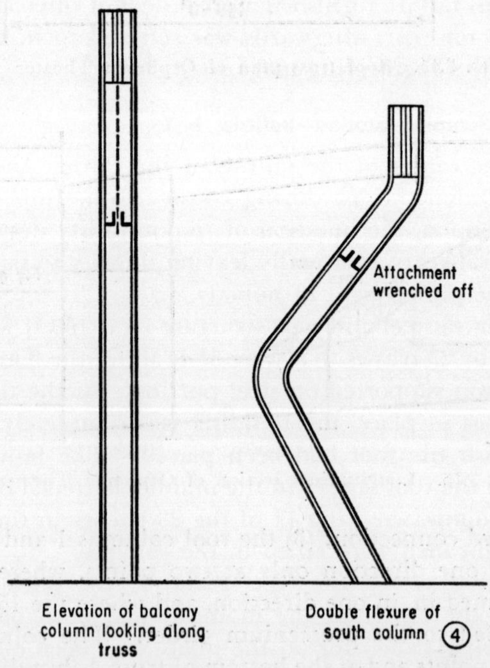

Fig. 3.18. Orpheum Theater column before and after failure

For some reason or other, the balcony was loaded with a much heavier load than had been figured, so that before the failure, under dead load only, some of the truss members were stressed to 22,000–24,000 psi and the bottom chord splice was stressed to 22,000 psi. The code at that time set a maximum stress of 16,000 psi. Under this excessive concrete load, the balcony columns were overloaded by about 50 per cent. The most critical part of the framing, however, seems to have been the long roof columns 1 and 2, and the questionable value of its lateral support. Depending on how much credit may be given to the two lateral supports noted above and shown on Fig. 3.14 and Fig. 3.15, the ratio of slenderness may be taken at either 156 or 283, instead of the allowable figure of 120, which is considered safe practice even today.

Some of the theories advanced for the failure were: (1) the torque of an already overloaded balcony truss caused by the twist of the front portion of the balcony cantilever trusses; (2) the excessive ratio of slenderness of the main roof columns; (3) failure of the balcony columns under the excessive and eccentric load of the concrete; and (4) a possible interrelationship of stresses developed in the finished structure that added the results of one weak point to another.

All in all, this failure furnished a great deal of interest for structural engineers, and for years afterwards was referred to in various articles and letters in the engineering magazines.

Porter Avenue Pumping Station—Buffalo, N.Y.

Although the causes of the failure of the Porter Avenue Pumping Station [50] in Buffalo, N.Y., during construction, June 30, 1911 were largely agreed upon, the question of responsibility dragged on in the courts for several years, eventually leaving the city to pay the bills and resulting in the punishment of nobody.

The building—a steel-mill-type structure—was 100 ft wide and 360 ft long, spanned by 23 trusses that were 53 ft above the floor. The cinder concrete roof was supported on steel purlins. At the time of the failure all steel was in place, the building was completely enclosed, and half the Spanish tile roof had been placed. The failure apparently occurred when the roof sagged in the middle, flattened out, and pulled in the side columns, wrecking 15 of the 23 trusses in the center of the building. Eight men were killed in the collapse.

Investigation disclosed the facts that: (1) the peak purlin of the trusses, under the center of the monitor, had been omitted; (2) the vertical X-bracing, called for on the plans, had been omitted; and (3) the truss members were $\frac{1}{16}$ to $\frac{1}{8}$ in. less in thickness than were required on the plans.

4

Miscellaneous Construction Failures

Most failures of buildings under construction, other than reinforced-concrete failures and the failure of structural steel frames during erection, are the failures of masonry walls. A few miscellaneous types of failures have occurred; a single example of several of these will indicate the hazards to avoid.

It is interesting to note that before the last decade of the nineteenth century, a large majority of the building failures would have come within the scope of this chapter. There are a number of construction accidents that are not really the type to include here—scaffold failures, trench cave-ins, and the like, and, in the course of the years, these have been accountable for a large proportion of the deaths in the industry.

4.1 Wind Failures of Masonry Walls

The author has been called upon to examine and report on literally dozens of failures of the kind described here—the failure of a partially constructed masonry wall during a heavy windstorm. These failures usually occur in bearing walls more than 14 ft high with comparatively few openings, such as those used in the construction of supermarkets, bowling alleys, school auditoriums, or gymnasiums. These failures, as a rule, happen when the wall has been erected but is without the benefit of the roof steel, bar joists, or the like to help support the wall against wind action as a vertical cantilever. Occasionally these failures occur after the wall has been stiffened by the roof construction, but this type is not nearly so common.

The contractor is ordinarily insured against wind damage, but he must take reasonable precautions, such as shoring the walls, until the support furnished by the roof structure permits him to remove the shores.

Before settling any claim for damage, the insurance company requires assurance that reasonable precautions were taken, and it was in the capacity of investigator and reporter that the author learned the facts on many of these failures.

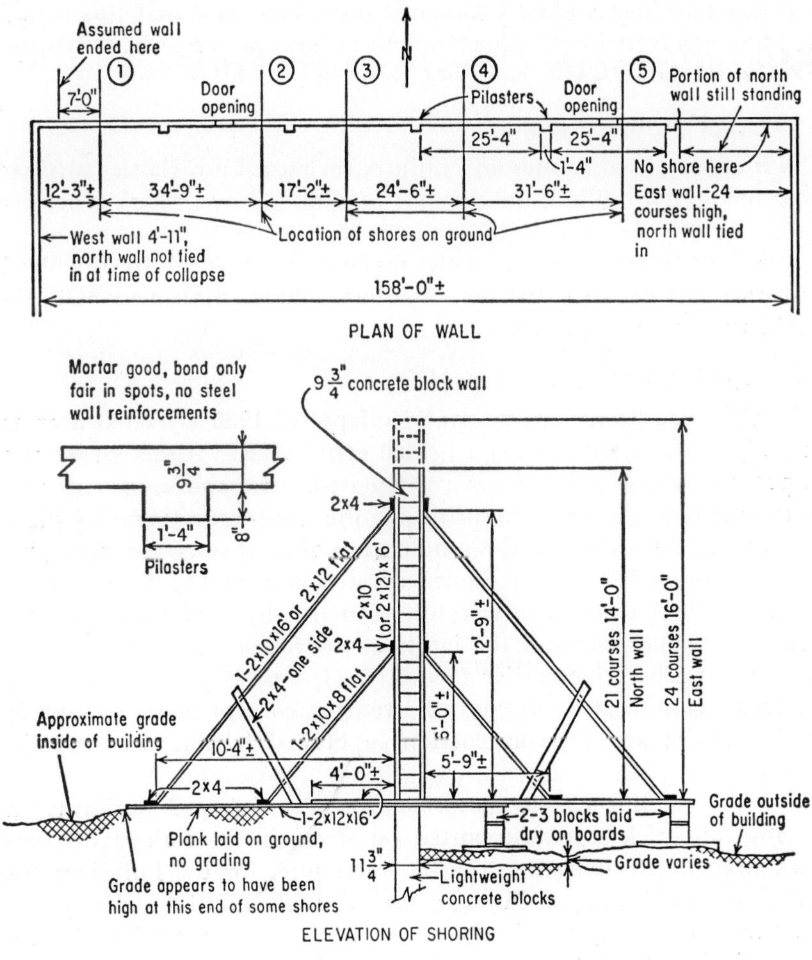

FIG. 4.1. Supermarket wall

Supermarket—Clarence, N.Y.

A typical masonry wall failure concerned the wall of a supermarket at Clarence, N.Y., which was 158 ft long, had been built of lightweight concrete blocks to a height of 14 ft, that is, 21 courses, and was braced as shown in Fig. 4.1. There were two door openings in the wall, making up about 2¼ percent of the wall area. The failure occurred during a windstorm on the morning of November 19, 1958, before any of the workmen were on the job. The entire length of the wall failed outward, wrecking all of the supporting shores.

It is probable that gusty winds, rather than the steady 32 mph wind that was blowing, gradually loosened the shores, and that this condition was aggravated by the small amount (only 2¼ per cent) of venting area in the wall.

4.2 Major Masonry Failures

A number of major masonry failures on record are classics of their type and do not fall into the category of wind failures, which generally do not involve death or injury and are of a relatively minor nature.

It is interesting to note that the majority of these failures involved apartment houses and that, in several instances, an element of negligence was definitely involved.

Bronx apartment collapse—Bronx, N.Y.

The famous Bronx apartment [51] collapse of 1936 attracted a great deal of attention and started a heated controversy. Although no one disputed how the collapse was precipitated, some controversy existed as to who was directly responsible for the misinterpretation of plans that caused the failure of the center wing of a six-story and basement E-shaped apartment house under construction in the Bronx, N.Y., June 19, 1936—a failure that resulted in the death of eighteen men. This failure had many of the faults of the famous Darlington Hotel failure of a generation earlier (Sec. 7.1): (1) a set of plans prepared by an architect who had no control over the construction; (2) shoddy workmanship; and (3) construction supervised by and done by the owners.

At the roof line of the center stem of the E-shaped building, an ornamental brick tower supported on steel girders a short distance back from the front wall housed a water tank for the building; the failure occurred under the bearing for this tank (Fig. 4.2 and 4.3).

As for the Darlington Hotel, the architect had been paid $1,200 for a set of plans for a $300,000 building. Unlike the Darlington, this was a wallbearing building with wood floors and joists throughout, designed to house 60 families. The owner and his son were the con-

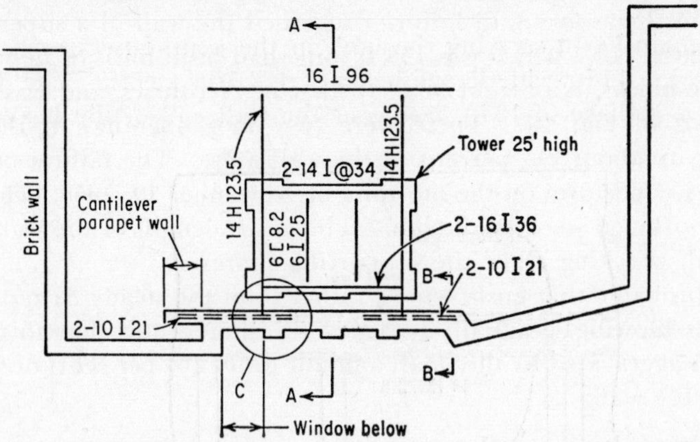

FIG. 4.2. Plan of center wing roof and tower framing of apartment house

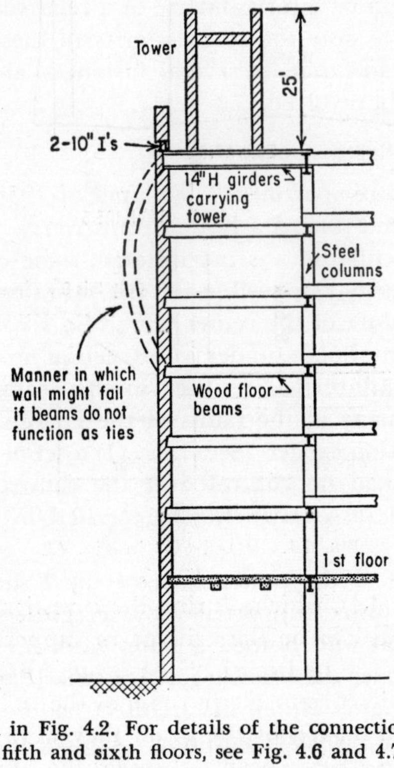

FIG. 4.3. Section A-A in Fig. 4.2. For details of the connections of the floor beams and outer wall at the fifth and sixth floors, see Fig. 4.6 and 4.7

tractors, and they bought the set of plans without any supervision by the architect.

The main beam carrying the tank in the penthouse area described above was a 14-in.-wide flange beam at 123.5 lb per ft. The drawing showed a grillage of 10-in. beams at the wall end of the heavy 14-in.

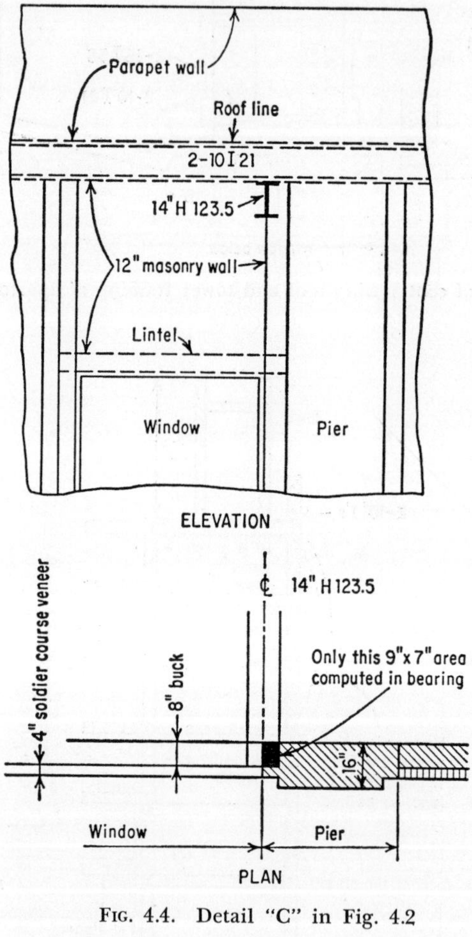

FIG. 4.4. Detail "C" in Fig. 4.2

beam (Fig. 4.2 and 4.4), which, during the investigation, the architect claimed was intended to be the grillage to support the 14-in. beam. However, the architect had marked the beam ABOVE, and, without consulting the architect as to the purpose of the 10-in. beams, the contractor placed them on top of the beam they were intended to carry (Fig. 4.5). Had the grillage been put under the 14-in. beam, it would have extended into the ceiling of the room below. The girders, as placed, were bearing partially over a window opening, and the brick wall was stressed to more than 1,000 psi—far beyond anything the New York code allowed on good brickwork. It was dry masonry with inadequate bond, low-grade brick, and excessively wide joints of very

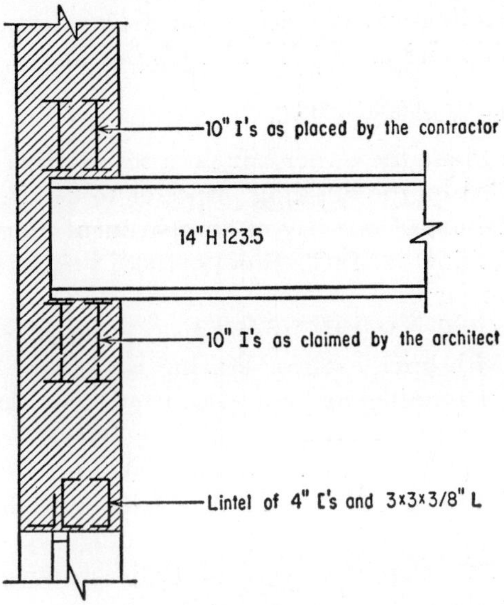

FIG. 4.5. Section B-B in Fig. 4.2

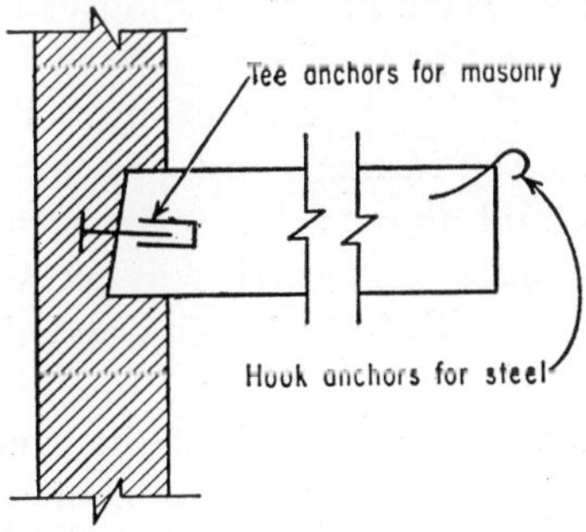

FIG. 4.6. Standard beam anchors used sparingly (see Fig. 4.3)

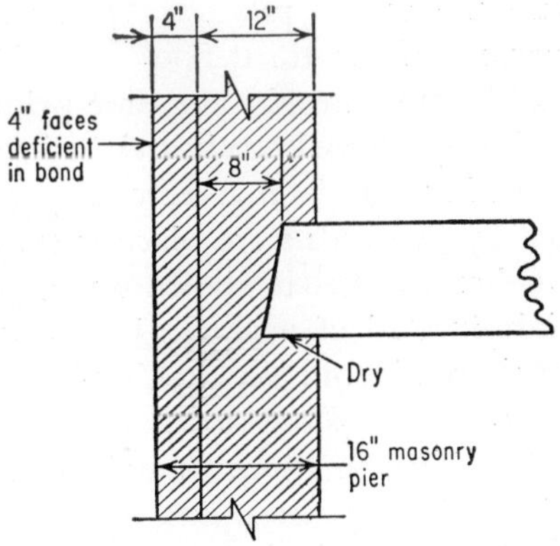

FIG. 4.7. Inadequate beam anchorage at a pier (see Fig. 4.3)

poor quality mortar. Joists and beams were inadequately anchored to the masonry throughout the whole building, and badly pitted second-hand steel was used in many places (Fig. 4.6 and 4.7).

The committee of experts who investigated the failure for the district attorney included, among others, two of New York's most prominent engineers, Elwyn E. Seelye and H. G. Balcom, and the report from the committee was an excellent one.

The district attorney stated, "I took a piece of mortar in my hand and it crumbled like dust. We found steel beams that had holes in them. The building was so defective in design that, even if 100 per cent workmanship and materials had been put into it, it had to fail."

Apartment house—Washington, D.C.

Another case where the owner bought a set of plans from the architect and depended on the building inspector to do all the supervisory work, was the failure of one bay of an apartment house [52] under construction in Washington, D.C., June 9, 1908, a failure which caused the death of two men.

The area that failed was the corner bay, about 12 × 16 ft, of a six-story building with brick exterior bearing walls, steel beams and columns for interior framing, and cinder-concrete floor slabs. Plans had originally been filed for a five-story building, but, after the lower part of the walls was built, it was decided to make it a six-story building. Although the building code required heavier walls for the six-story building, the permit was issued by the building department without any change. The collapse occurred while the workmen were placing the concrete for the corner bay on the fourth floor.

The failure involved the third and fourth floors, the adjacent wall, and the corner pilaster. The shores for the fourth floor were supported on the third floor, but the shores under the third floor slab had been removed some time prior to the accident. Inspection of the building showed cracked slabs in other exterior wallbearing bays, but none in the slabs that were supported on all sides by steel beams. The reinforcing rods were placed very irregularly, in many instances much further apart than shown on the plans, and about 2 in. up in a 4-in. slab, instead of near the bottom. The slabs were carried in recesses in the walls, but the depth of the slot was as little as 1½ in., and the concrete was poorly tamped into the slots.

The coroner's jury brought in a verdict to the effect that "the cause of the fall we determine to have been the faulty construction of the corner pier, which was not of sufficient strength as computed and passed by the building inspector's office."

1078 Madison Avenue—New York, N.Y.

An apartment house [53] under construction at 1078 Madison Avenue, New York, N.Y., collapsed September 28, 1896, resulting in the death of one man and injury to eight others.

The building was 25 × 85 ft in area. The walls had been completed up to the third floor, beams were in place, and the bricklayers were laying walls above the third floor when the failure occurred without any warning. An investigation disclosed that the construction was faulty throughout: (1) the foundations were insufficient and had cracked before the failure; (2) the contractor had skimped on wall thickness; (3) the bricks were badly laid in poor mortar; (4) the lintels were not of sufficient strength; and (5) the chimney flues instead of being lined throughout were only lined at points that might be open to inspection. It developed that the building department was so undermanned that 57 inspectors had to follow 4,200 jobs; the inspector who covered this area had never been on the job.

The following quotation from the coroner's jury's report has a familiar tone:

> We hold responsible the Department of Buildings of the City of New York for insufficient inspection. We deplore that a lack of finances tie the hands of the Department of Buildings while the law impels it to perform inspections which, under the present conditions, are practically a physical impossibility to the available force.

Saxe Theater—Janesville, Wis.

The failure of the new Saxe Theater [54] at Janesville, Wis., September 18, 1923, resulted in the death of one man. There were many indications of negligence in various portions of the work. In the usual attempt to cut costs there was a sad division of authority between the owner, the architect, and the various contractors. The architect sold a set of plans without any structural design and without any responsibility for inspection. The steel design was made by the steel fabricator, but the steel was sold to the general contractor to be erected by him. The coroner's jury placed the responsibility for the failure on the design, the construction, and the inspection. The insufficiency of the state building code and of the inspection by the state and by the city of Janesville also came in for criticism in the report of the state department of engineering. One point seems to be definitely agreed upon, however, that the precipitating cause of the collapse was the failure of a masonry pier; for this reason, the author has chosen to include it here.

The main portion of the building was 60 ft wide and 120 ft long, with 12-in. walls, 28 ft high. It was spanned by five trusses spaced about 22 ft apart and carried by brick piers so that the total size of piers at these points was 17 in. thick by 29 in. long. The piers were shown on the plans to be solid brick to within 2 ft of the top—the upper 2 ft consisted of poured concrete at the end of the trusses.

All the piers and bearing plates were to be the same size, although one end of the southernmost of the five trusses carried a much heavier load because it supported the penthouse and ventilating machinery (Fig. 4.8). The trusses, of about 60-ft span, were about 6½ ft deep at

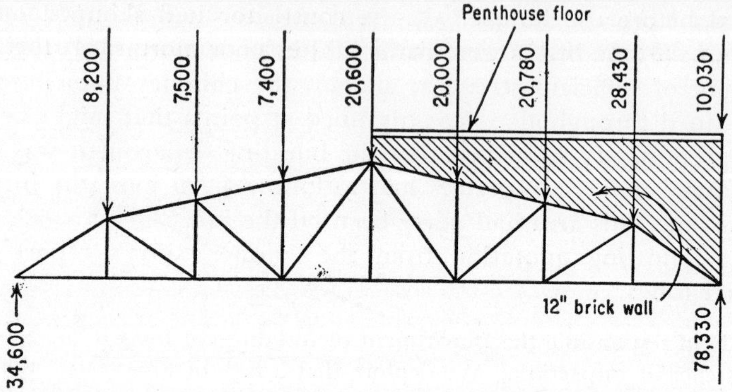

FIG. 4.8. Dead load on south truss carrying penthouse of Saxe Theater

the center with sloping top chord, tapering down to 12 in. at the ends (Fig. 4.9). The north penthouse wall, shown on the drawings as a 9-in. wall, was actually built as a 12-in. wall and was carried directly on this sloping top chord with no clips to prevent the sliding.

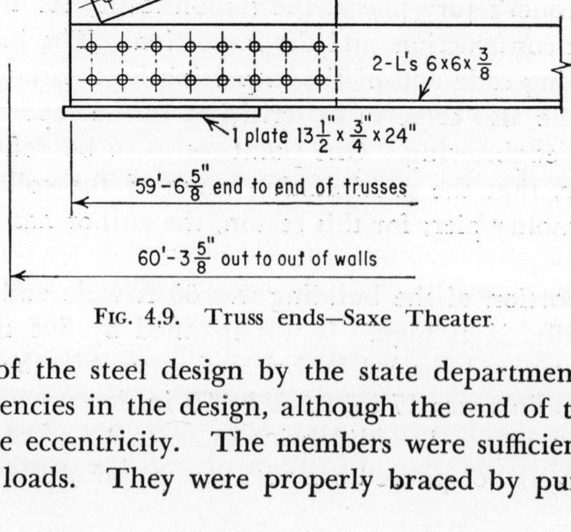

FIG. 4.9. Truss ends—Saxe Theater

A review of the steel design by the state department revealed no serious deficiencies in the design, although the end of the truss (Fig. 4.9) had some eccentricity. The members were sufficient to carry all their design loads. They were properly braced by purlins at panel points and had satisfactory cross bracing. The 3½-in.-thick concrete roof slab also was of satisfactory design.

The building was about completed when the failure occurred. The entire roof failed, pushing out one side wall completely, and part of the other wall, but the stage end and front wall remained upright. The roof trusses, although badly damaged by the fall, showed no evidence of damage due to overstress. The connections and rivets were

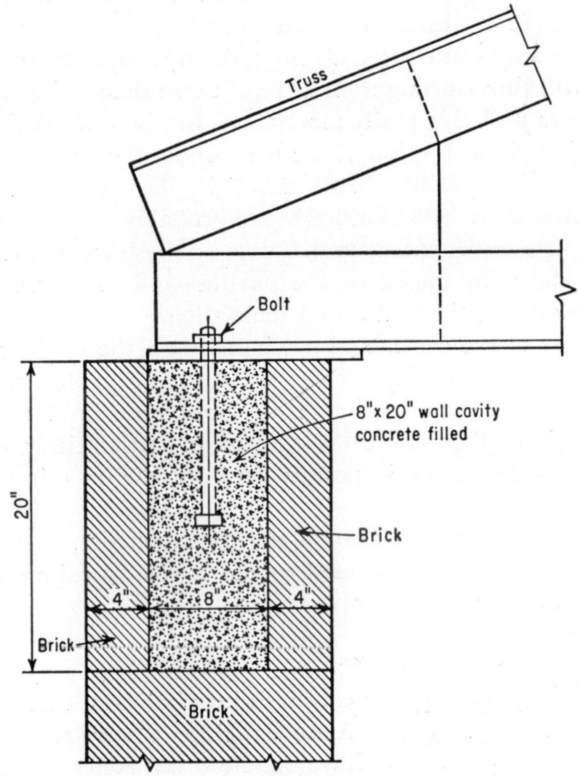

FIG. 4.10. Bearing piers—Saxe Theater

still intact after the fall. The concrete, although broken up by the collapse, showed remarkably good strength.

Subsequent investigation disclosed that the weakness lay in the manner in which the construction of the bearing portion at the top of the piers was changed. Whereas the design called for the bearing plate to rest on concrete over its entire area, the construction had been changed so that the top of the pier had been built as a brick box into which a concrete pier 8 × 20 in. and about 20 in. deep had been poured (Fig. 4.10). When the truss had been placed, the anchor bolts were dropped into the brick box and the concrete poured in to fill the cavity and tie the anchor.

Although the bearing plates were all the same size, the end of the south truss, where failure began, carried about double the load of any of the other bearing points because of the load of the penthouse at this point—a total dead load of almost 80,000 lb. This load, spread uniformly over the bearing plates, would have loaded the pier to about 250 psi, which would not have been excessive. However, because of the difference in hardness and in modulus of elasticity of the brick and concrete, the load was carried eccentrically by a concrete area of 8 × 20 in. (Fig. 4.10) thus causing a pressure of more than twice this amount. The load was probably really more than this, because of the eccentricity of bearing, the deflection of the truss under the heavy weight of the penthouse, and a possible thrust caused by the method of supporting the penthouse wall. The failure began probably through the crushing of the concrete under this truss bearing; as the truss fell, it pulled the next truss down by means of the purlins—and so on until, through progressive failure, the entire roof had fallen.

The report of the coroner's jury, based on a thorough examination of the ruins and on the reports of a number of capable experts, was to the effect that: (1) the architect's drawings were insufficient; (2) the building code of the state of Wisconsin was insufficient; and (3) the inspection by the state inspectors and by the city inspectors was insufficient.

It is interesting to note that in the reconstruction of the building the original steel plans were used except that steel columns were substituted for masonry piers.

Alameda Hotel—Kansas City, Mo.

Although the report of the coroner's jury indicates that the collapse, during construction, of the Alameda Hotel [55] in Kansas City, Mo., September 26, 1912, was a failure of reinforced concrete, there can be little doubt that the actual cause of failure lay in the masonry instead. Four men were killed and others injured. The report of the coroner's jury does not follow the usual pattern of such a report, but points successively to the administrative errors which permitted it to happen. In this sense, it is a document well worth reading. It is as follows:

> We find that the cause of the collapse of said Alameda Hotel building was due to the following successive causes:
>
> That the owner of said Alameda Hotel building used an improper method of obtaining the reinforced concrete design, to wit, by letting the contractor for the reinforced concrete and the materialmen furnish this design, thereby making it an inducement to the contractor and to those who had the

subcontracts under him, to make a design embodying a less amount of material than is specified by the building code.

That the design of the reinforced concrete made by the Trussed Concrete Steel Co., of Detroit, Mich., is weak and in many respects not in accordance with the building code.

That the architect of the building to whom this reinforced concrete design was submitted, passed it on to the contractor without checking the design.

That the superintendent of buildings issued the permit for this building without checking the plans and ascertaining that the reinforced concrete design was weak and not in accordance with the building code.

That the building was not properly inspected during the course of erection by the department of buildings due to lack of a sufficient number of inspectors in this department.

That the owner in undertaking to superintend the erection of the building himself assumed full responsibility for the actions of the workmen, and that it was his duty to see that the forms were left under the concrete for a proper length of time and that the plans and specifications were carried out, which he did not do.

That the contractor for the reinforced concrete placed a weight of cinders on the roof and pulled the forms in the presence of the owners while the concrete was still weak and not capable of supporting its own weight and the superimposed load, and this without having made a sufficient examination of the concrete to ascertain its strength, and that the roof slab failed under the superimposed load and carried the other floors and portions of the walls down.

And we are further of the opinion that if the design for the reinforced concrete had been made in accordance with the code, that the roof slab probably would have stood up after the forms were removed: that the use of lime mortar in the brick walls, and insufficient bonding probably did augment the disaster.

At no place in the report is there any clear-cut statement as to what caused the failure—only that the shores were removed too soon. However, Mr. C. A. P. Turner, one of America's foremost engineers, was on the job shortly after the collapse, and his letter in the *Engineering News* is much more specific: (1) the concrete was first-class; (2) the bearing of the slab on the brick walls throughout was only on the inside 4 in. of brickwork; (3) the 4-story bearing wall was only a 12-in. wall throughout its height; (4) the 4-in. facing and the 8-in. backup were insufficiently bonded, thus giving the wall the effect of being an 8-in. wall only; and (5) lime mortar was used throughout.

Brick wall—Louisville, Ky.

Operations performed out of sequence can bring about a failure and have frequently done so. In the case of the failure of a 25-ft-high brick wall [56] under construction in Louisville, Ky., February 13, 1914, four workmen were injured when a wall, which was weakened by a recess at its base, failed. The recess was intended to receive a concrete slab to be poured at a later date.

Frozen mortar failure—St. Louis, Mo.

The mortar in a masonry wall is as vulnerable to the hazards of winter construction as concrete. If the water content of the mortar freezes shortly after its initial set has taken place, the initial set is destroyed by the expansive action of freezing, and the mortar will never attain its full strength. A news item in the St. Louis, Mo., papers called attention to the collapse, during the early morning hours of March 31, 1912, of a two-story brick building.[57]

The mortar was mixed when the temperature was 16°F. Without knowing all the details it would be safe to assume that the mixing water must have been heated; otherwise, exposed to this temperature, the mortar would have been solid ice before the bricklayer had been able to lay the wall, and there would have been no bond whatsoever. The fact that the building stood until it was almost completed would indicate that the mortar actually did have some small amount of strength. Otherwise it would have been crushed by the dead weight of the wall as it was laid up.

4.3 Timber Roof Truss Failures

Timber roof trusses require the same care in erection, the same adequate bracing, that steel trusses require, and lack of this bracing will permit their failure in the same manner.

Safeway Stores—Redwood City, Calif.

Seven timber trusses, 105 ft long, in series, fell when a building [58] for Safeway Stores at Redwood City, Calif., failed in May, 1954, injuring six men.

The trusses were of the bowstring type with a top chord made of two 3 × 14 members joined to make a segmental arch and with the compression splices in the two members staggered. The bottom chord was composed of two 3 × 12 timbers with tensile splices. Permanent longitudinal bracing of two lines of 6 × 6 struts with 2 × 6 boards as cross bracing was located at third points in alternate bays only—forming a box structure of each pair of trusses. Some 2 × 12's were nailed flatwise as temporary bracing in the open bays. The 2 × 12 roof and ceiling joists were piled on the roof and some of them had been installed.

Two of the theories of the cause of the failure were: (1) the heavy pile of roof and ceiling joists was concentrated so that one or more of the trusses were overloaded at a critical point; and (2) the failure was caused by the settlement of a temporary post that supported one end of a truss until the wall was built to receive it, the bottom of the post resting in a soft, wet spot. Certainly a contributing factor was the fact that neither of the end walls nor part of the side walls had yet been erected, so that some of the trusses lacked sufficient lateral bracing. The primary failure was triggered by the lateral buckling of the top chord of a truss at a splice point, thus kicking out the bottom chord laterally in the opposite direction.

4.4 Church Floor Failure—Harrington Park, N.J.

Regardless of the fact that it is rather difficult to understand the sequence of construction indicated by the statement of the contractor during the investigation, one thing is evident in the failure of a church floor [59] at Harrington Park, N.J., during the dedication services, April 14, 1912. Those in charge of the affair—the church board, the building committee, or the pastor—had set April 14 as the day of dedication, and come heaven or high water, finished or unfinished, the church was going to be dedicated on that day.

Approximately 300 people were gathered on a 40 × 60-ft floor space when the floor gave way down the center and precipitated the entire group into the trough thus formed, killing 2 people and injuring about 50 others. Piles of loose lumber stacked along the sides of the room slid into the mass of people, compounding the seriousness of the collapse.

The contractor claimed that the floor was supported by temporary supports only and that no permission had been granted for holding services in the church. It is strange that "temporary supports" would be used to sustain the floor until the building had reached the stage where interior finish was being installed.

4.5 Cave-in

Hotel Samuels—Jamestown, N.Y.

A typical cave-in type of failure, such as is discussed more fully in Sec. 6.2, "Removal of Lateral Support," was the collapse of a three-story portion of the Hotel Samuels in Jamestown, N.Y. shortly before noon, December 29, 1949. Fortunately, no lives were lost.

The hotel was built in 1910. The ground floor, which was occupied by a very busy restaurant, was made up of a concrete slab on grade.

The second and third floors and roof were of concrete on steel beams, supported on steel columns in the interior and the wall bearing at the exterior ends. The wall was a 2-ft-thick, solid brick wall up to the second floor, and 1-ft 4-in. brick above this to the third floor, then dropping back to 1 ft for the remainder of the height.

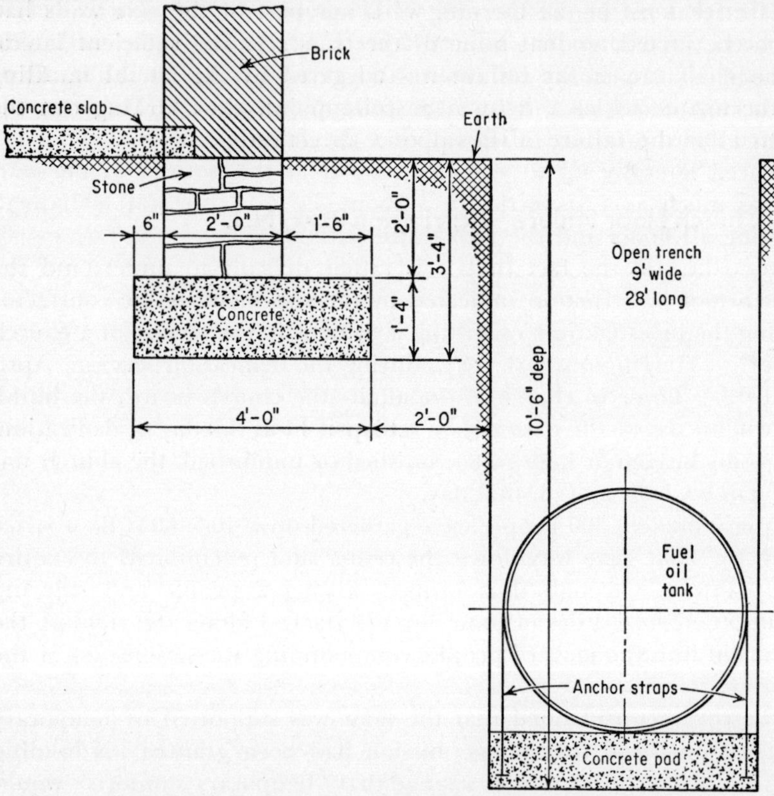

Fig. 4.11. Detail of excavation alongside Hotel Samuels

No plans of the building were on file with the building department, but measurements on the site indicated that the brick wall had been built on a 2-ft-thick stone wall, 2 ft high, which in turn rested on a 4-ft-wide, 1-ft 4-in. deep unreinforced-concrete footing which was so placed that it had a projection of 1 ft 6 in. on the outside and 6 in. on the inside. Since stone foundation walls were not being built in 1910, it is probable that an existing foundation wall had been used for the new construction. The foundation material was a yellow-brown clay.

An excavation 28 ft long, 9 ft wide, and 10 ft 6 in. deep for a fuel storage tank had been made parallel with the building wall and 2 ft from the footing (Fig. 4.11). The concrete for the anchor pad had been poured and anchor straps placed. At the time of the failure, there probably was a difference in elevation of between 5½ and 6 ft from the top of the new concrete pad to the underside of the footing and only 2 ft from the edge of the footing to the side of the excavation.

Rain fell most of the morning of December 29 and, shortly before noon, the superintendent noticed a crack appearing in the wall. He immediately ran to the restaurant and gave warning, enabling all of the occupants to leave before the collapse occurred. Measurements taken after the failure disclosed that the foundation wall had been displaced laterally up to as much as 5 ft at the bottom of the wall and as much as 4 ft vertically. About 35 ft of the wall collapsed, carrying all floors and the roof down with it.

5
Failures Caused by Alterations

The comparatively few cases in this chapter are typical of the hazards of alteration construction work; they should impress anyone concerned with this type of operation with the necessity for complete knowledge of the dangers that may be encountered.

5.1 Failures as a Result of Alteration Operations

In many old buildings previous alterations may have been made and never recorded—alterations that may have weakened some part of the structure. Frequently, alterations may be made by some employee of the owner without the advice of an architect or engineer, perhaps by a handyman not familiar with possible dangerous conditions that may have been or may be introduced by the operation proposed.

Even if the operation has been planned by an architect or engineer, he may assume conditions that do not exist. For example, he may assume that proper bridging or wall ties were built in, on the grounds that today's building code would require them, or he may see what appears to be a pilaster in a brick wall, only to find out later that it is an unused flue, incapable of carrying any load.

Usually, alterations are carried on under crowded conditions, and storage space for materials is at a premium. As a result, bags of plaster, piles of tile, lumber, or other construction materials are piled on floors incapable of carrying such loads. In most of the failures described here, each of these factors is present at some point.

Elks Club—Los Angeles, Calif.

Frequently, in an existing building, what is visible does not tell the real story of what is beneath the surface. There are times also when what appears to be a simple operation should not be undertaken without professional advice—particularly if construction plans of the building are not available. In the case of the Elks Club [60] of Los Angeles, Calif., the simple operation, in 1929, of cutting a new door through an existing wall, brought about an incipient failure, necessitating emergency and permanent repairs, which brought the cost of the door to $10,000.

For the sake of convenience a doorway was to be cut through what appeared to be a simple partition. The portion of this partition through which the door was cut, however, turned out to be a reinforced-concrete girder of 60-ft span, 15 ft deep. When serious cracks appeared, the engineers were called in and the true condition was discovered. Three 5/8-in. vertical stirrups had been cut, and the main reinforcement in the bottom of the girder was exposed at the door threshold. The girder, which was 18 in. wide, was carrying a computed maximum moment of 4,750,000 ft-lb and a shear of 314,000 lb at the support, about 85 per cent of which was dead-load moment and shear. By cutting the doorway, the average resultant shearing stress, without the stirrups, was 175 psi, several times the safe stress.

It is probable that complete failure was prevented by three factors: (1) the supporting effect of a 6-in. hollow-tile wall, which relieved part of the load; (2) the effect of fifteen 1¼-in. square bars in the bottom of the girder; and (3) relief of part of the load by the cantilever effect of the roof joists. Temporary steel shoring was immediately placed, and permanent repairs were accomplished by installing two parallel steel trusses, one on either side of and adjacent to the failed girder.

Kimberly-Clark Paper Co.—Kimberly, Wis.

Another case of the unwarranted assumption of structural conditions that did not exist—but this time with the supervision of competent engineers—is brought out in the collapse of a two-story section about $54 \times 75\frac{1}{2}$ ft in area of a building [61] in the plant of the Kimberly-Clark Paper Co. at Kimberly, Wis., in 1927.

The original building, of brick and stone masonry construction, had been built in 1890; in the spring of 1926, the entire interior was rebuilt, using steel framework and reinforced-concrete floors. The only parts of the old building left intact were the four brick walls, and the two stone flume walls in the basement that formed two bearing walls, thus making three equal spans for the floors above (Fig. 5.1).

The timber flumes against the flume walls were also left in place. Because of the water in the flumes, the arches in the masonry wall were not visible, and it was assumed that these walls were solid so that the columns for the reconstruction could be placed anywhere. Investigation following the collapse showed that neither of the two col-

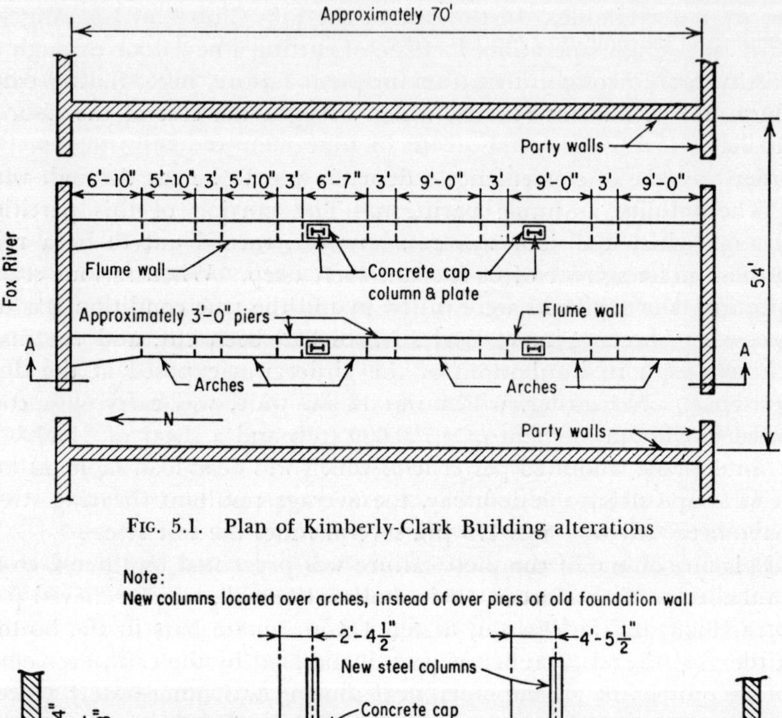

FIG. 5.1. Plan of Kimberly-Clark Building alterations

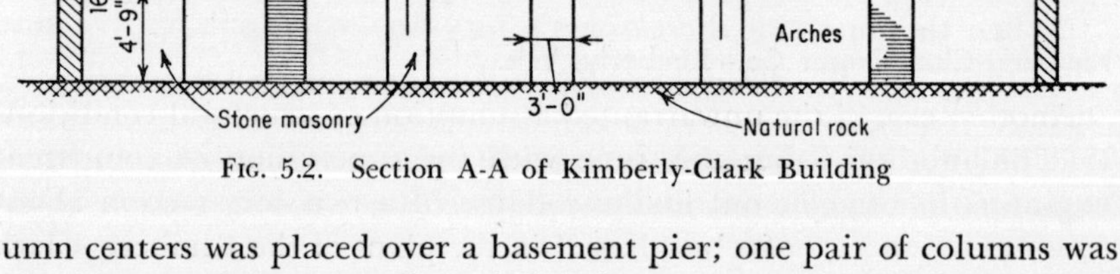

FIG. 5.2. Section A-A of Kimberly-Clark Building

umn centers was placed over a basement pier; one pair of columns was 2 ft 4½ in. off the centerline of the pier, or actually 10½ in. out over the arch. The other pair was even worse, 4 ft 2½ in. off the center of the pier, or out onto the arch 2 ft 11½ in. (Fig. 5.2). Workmen were removing the old timber of the flume when the collapse took place.

At the time of the failure, the first and second floors were loaded to 192 and 252 psf, respectively—on floors that were designed for 250 and

350 psf, so it is obvious that the floors were not overloaded. A possibility exists that the removal of the wooden flume may have disturbed the equilibrium of the column loads or removed a small amount of support from the piers, but it was quite evident that the main cause of the collapse was the crushing of the old arches under the load of the new steel columns.

Building at Eleventh and Market Streets—Philadelphia, Pa.

A typical collapse during alterations took place July 15, 1909, when about half of a 50-year-old building [62] in Philadelphia, Pa., 20 × 50 ft in size, failed, causing the death of nine men and injuring fourteen others.

The building, of brick bearing wall and timber frame construction, was originally a three-story structure, but had been extended to five stories. The second floor was to be removed to make one large exhibition floor the full height of the two stories. This work was under way when the failure occurred. Needle beams made up of 8-in. steel I's were supported on 8 × 8 posts, 36 ft long, spaced from 2 ft 6 in. to 4 ft apart, and bearing on 12-in. wood sills on the sidewalk and on the basement floor.

Since those who knew what happened were all killed in the failure, it was impossible to determine the exact cause of the accident. It may have been any of the following or a combination of several: (1) the displacing of one of the shores by a steel girder that was being swung into the building at about the time of the collapse; (2) the attaching of a block and fall to one of the sills or shores to aid in placing a girder; (3) the removal of some of the shoring to allow for the passage of the girder; or (4) the collapse of the fourth floor under the excessive load of a block and tackle that had been placed there to raise the girder.

When the report of the coroner's jury was made public, it blamed the failure on "the pulling out of joists by vibration, due to concealed defects."

The building was found to be much more unstable than superficial examination had shown, and any remodeling was obviously a hazardous operation.

Strathmore Apartments—New York, N.Y.

The failure of a section of the Strathmore Apartments [63] in New York, N.Y., December 1, 1920, might have been placed more properly in Chap. 9, because most of the causes of this failure as determined by investigation are in the list given in Sec. 9.1. Nevertheless, the collapse occurred during alterations, and there seems to be no doubt that insertion of needling, thus removing a small amount of stiffness was the final straw, even though it was not in the area where the actual failure occurred.

The building was a nine-story brick wallbearing apartment house, 70 × 62 ft, built in 1880 to accommodate sixteen families. The walls were of brick with lime mortar and, in a number of places, were undoubtedly weakened by the presence of old flues, which were installed before the days of central heating. These flues can sometimes give the false impression of greater safety, since a later generation of builders may assume that the added wall thickness required for such flues indicates the presence of pilasters, instead of being a plane of weakness. Interior brick walls had originally carried the 3 × 8 floor joists, which were spaced unusually close together—from 3½ to 5 in. on center for a span of 18 ft. At the east wall where the failure began, these joists were carried by brick corbels, extending a very short distance into the wall and with no positive anchorage to the wall. During the 40 years since it had been built, the building had undergone numerous previous alterations—notably the substitution of steel framing in the two lower floors in place of the original wall bearing.

At the time of the failure, further extensive alterations were in progress to permit the use of the two lower floors for commercial purposes. These alterations consisted primarily of the removal of brick bearing walls and the substitution of steel framing in the lower stories, for both the exterior and the interior walls. Most of the work being done was at the west end of the building toward Broadway where the brick walls had been supported on needle beams at the fourth floor level, and supporting girders and columns were being placed. While this work was underway, a gradual failure of the walls toward the southeast corner, the opposite end, gave warning so that most workmen escaped; however, one man was killed and about twelve men were injured when the collapse occurred. Several persons in an automobile salesroom on the ground floor of the corner escaped from the building. Several hours later, while some loose portions of the wall were being pulled down, a further area collapsed. As a result of the two failures, the entire south half of the east wall and the east half of the south wall and an interior bearing wall parallel with the east wall failed, carrying down all floors bearing on them. The exterior wall failures occurred in an almost vertical line through lines of windows in both of the walls.

The remainder of the structure apparently suffered little damage. The Strathmore was separated from an adjoining five-story apartment by a court about 4 or 5 ft wide; this court was filled almost to the second floor window level, with the debris from the collapse, but little damage was done to the building. The northern half of the street was filled with debris.

Investigation by the New York Building Department, under its nationally known and respected superintendent, Rudolph P. Miller, blamed the collapse on a combination of circumstances—practically absolving the present alterations from any blame. The causes listed included: (1) lack of proper anchorage between the floor and wall construction throughout the building; (2) weakened masonry, disclosed when the plaster was removed, indicated by numerous cracks in the brick walls; (3) poor bond in the brickwork; and (4) increasingly heavy vibration caused by installation of a subway, streetcar traffic, and increasingly heavy truck traffic.

Although probably no nine-story wallbearing building would be built nowadays—for economic reasons, if for no other—the lesson to be learned if we are to profit by it is that, in any old building of this type, if any alterations are undertaken, the brickwork is immediately suspect. A thorough investigation of everything pertaining to the masonry wall, both for lateral and vertical strength, should be undertaken. The "good solid construction of the old days" may not be as good as you think.

Temporary shoring—Rochester, N.Y.

Shoring operations are always hazardous, and require more care and supervision than other types of construction. A building [64] undergoing alterations in Rochester, N.Y., failed, December 2, 1912. The second floor beams were supported temporarily by shores resting on sills on a series of unbridged joists. This procedure in itself was unsafe. However, nothing failed until the second floor was overloaded by a pile of construction materials. The unbridged joists tipped and failed, dropping the second floor beams and killing four workmen who were in the cellar at the time of the failure.

5.2 Failures Due to Previous Alterations

Not all failures due to alterations occur when the work is done. Sometimes an unsafe condition may be built into a structure and remain until some later date when an overload or other cause will bring about a failure.

Associated American Household Warehouse—Buffalo, N.Y.

About November 20, 1957, an area 18 $\times$ 22 ft of the second floor of a warehouse in Buffalo, N.Y., failed. No injuries occurred and no other part of the building failed except this one, on which there was a total superimposed load of about 20,000 lb.

The building was a well-constructed brick building, with wood flooring on wood joists, both in good condition. The load was not excessive for the floor as it had originally been built, but at some time

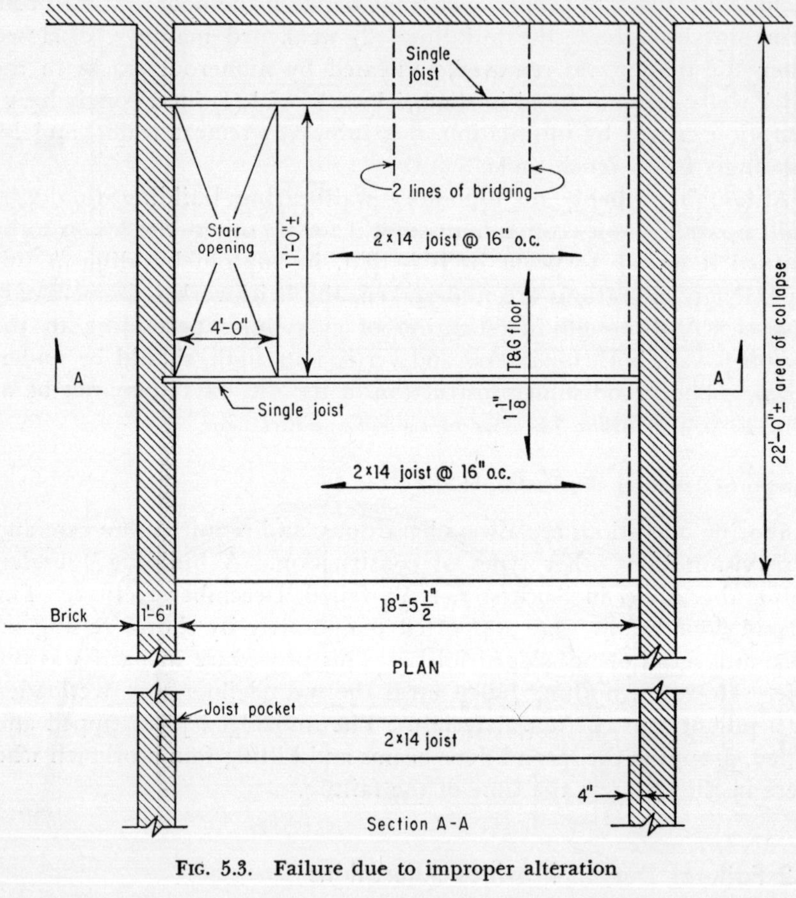

FIG. 5.3. Failure due to improper alteration

during the life of the building, a stairway from the first to the second floor had been cut into the second floor next to the wall, carrying the shortened joists on a single header joist, which in turn was carried by single joists (Fig. 5.3). Computations for the joists at either end of the stairway show that for a 120-psf floor load, which is the required load, this joist should be a triple joist instead of a single joist. The failure of these two joists in bending precipitated the failure, pulling adjacent joists with it as indicated in the sketch.

Olney and Carpenter quonset hut—Elma, N.Y.

Another failure resulting from alterations made at an earlier date was that of the quonset hut belonging to Olney and Carpenter at Elma, N.Y.

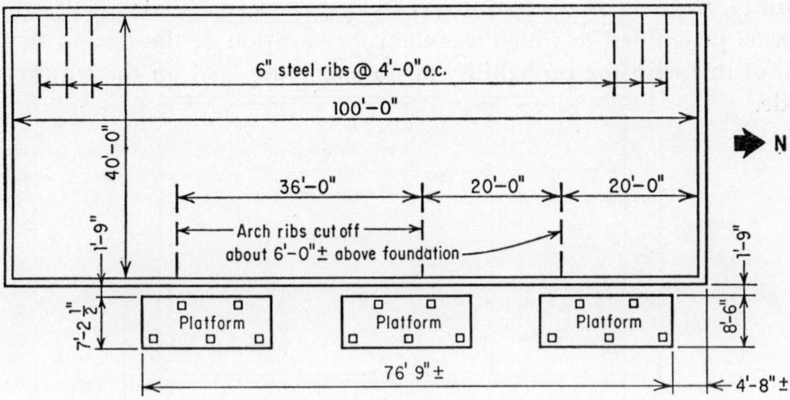

FIG. 5.4. Plan of quonset hut alterations

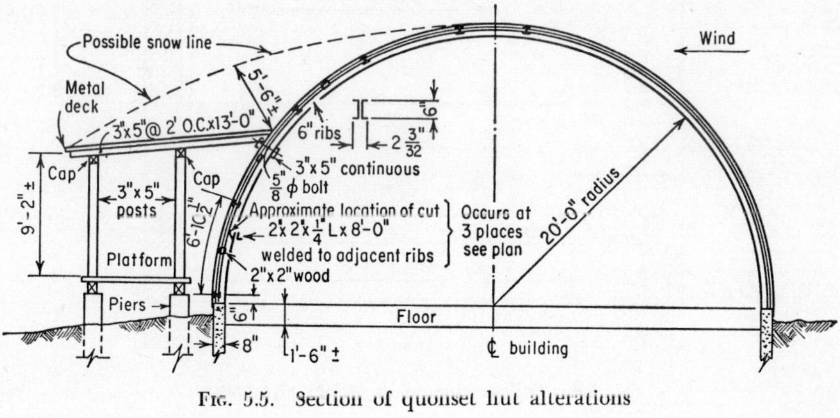

FIG. 5.5. Section of quonset hut alterations

This is the failure of a steel arch structure—a quonset hut under a heavy eccentric concentrated load (Sec. 1.4). The building, built in 1946 as a standard 40-ft semicircular arch quonset hut, 100 ft long, was later altered by the addition of a lean-to (Fig. 5.4 and 5.5). The metal roof of the lean-to was supported on wood rafters as a concentrated load on the arches. Three of the arch ribs had been cut and the lower sections removed to permit the installation of conveyors. The load from the cut arches was carried to adjacent ribs by means of steel angles spanning 8 ft. The ribs which had been cut were not adjacent, but were some distance apart.

A 36-hour snowstorm had piled 21 to 36 in. of heavy wet snow on the roof; wind drifting resulted in a possible 5 ft 6 in. of snow on the leeward side at a point where the lean-to was carried on the arch ribs. During the night of November 23, 1957, the roof failed. In all probability, the load at one point was 100 psf on a roof that was designed for 30 psf—added to which was the concentration of the lean-to roof—all of this with the probability of no balancing load on the windward side.

6

Foundation Failures

Foundation failures of some kind are much more common than may be gathered from the sections in this chapter. Most of them do not develop into dangerous conditions; however they cause unsightly cracks and depreciation of the building. In a great many low-cost residences, concrete block walls are laid up for the basement, backfill is placed, the house is finished, and sold to an unsuspecting owner. Then a few years later, a long horizontal crack appears somewhere in the middle third of the wall along a mortar joint. The wall is actually a vertical beam supporting the lateral thrust of the earth and, turned sidewise, it is undergoing the same type of bending as indicated in Fig. 1.2, with the basement floor and the first floor joists acting as supports and the earth pressure forming the load. The inner face of the concrete block wall is in tension, the mortar is incapable of taking this stress, and it cracks. Hundreds of these failures take place annually, and they are discussed in Sec. 11.15 under the subject of "Minor Failures."

Until recently, the determination of the bearing capacity of the soil was made by the engineer after a visual inspection, aided perhaps by some knowledge of how neighboring buildings on the same subsoil had been standing up. Then someone conceived the idea of making load tests of the soil and observing how the soil compressed under a load of one and one half times the soil pressure he proposed to use. From that the subject of soil mechanics was developed, by means of which the study of the behavior of soils became an exact science, with the result that nowadays few major jobs are undertaken without the services of a soil technologist. Many of the failures in this chapter would not have occurred had such services been available when the work was undertaken.

Uniform settlement of a building due to compressibility of the subsoil may not be serious. If the settlement under different parts of the building varies, however, or if the subsoil flows out from under the building, trouble may ensue.

The failures in this chapter are those in which there is little doubt that the failure of the foundation was the real cause of the trouble, even though other contributing causes may have been present. A review of other chapters in this book, however, will demonstrate that foundation trouble was a contributing factor in many of the other failures, notably formwork failures, temporary grandstand failures, and the failures given in Secs. 3.5, 4.2, 4.4, 7.1, and 9.3.

6.1 The Leaning Tower of Pisa

Most building failures are economic headaches, but this one has paid dividends throughout the centuries. In any chapter on foundation failures, the Leaning Tower of Pisa [65]—the grand-daddy of them all—should have first place.

The Tower was begun in 1155 to compete with the bell tower of St. Mark's in Venice. Before it had gone very far however, even before it was 40 ft high, it began to tilt noticeably. Founded on a deep saturated volcanic ash, the subsoil was evidently of uneven character over the site. The architect attempted to correct the condition by building the first, second, and third colonnades more nearly perpendicular, but without success. The tilting continued to increase. Whether by request, or of his own volition, he left the job, discouraged. Sixty years later another architect took over, but, after finishing the fourth arcade, he too quit. A third architect, William of Innsbruck, attempted to rectify the condition by building the fifth and sixth arcades with columns of varying lengths, the longer columns on the low side. The settlement still continued, and he gave up in despair. Almost one hundred years later, Tommaso Pisano, the fourth architect, completed the sixth balcony and built the belfry slightly offset, that is, more nearly vertical. The building was finally finished in 1350, almost 200 years after it was started.

Through the centuries various attempts have been made to correct the condition, but so far the successful and acceptable solution has not been found. In 1959 the tower was 17 ft 2½ in. out of plumb in a height of 179 ft, and the tilt is increasing, although not at a uniform rate. In the past 50 years, it has tilted 8½ in., about double the rate of the preceding 80 years. The increase in acceleration in 1958 was 0.046 in., the greatest in any year so far—so it appears to be gaining in momentum. Engineers' estimates give the tower a prospective life of 50 to 300 years, but this acceleration may alter the timetable.

In the meantime, it continues to attract tourists at the rate of 350,000 per year. With better engineering know-how and experience each year, it may probably be preserved so that it will not collapse and can continue to yield this rich harvest for centuries to come.

6.2 Removal of Lateral Support

The greatest number of serious foundation failures occur as the result of excavations that remove the lateral support of the subsoil under the foundation of a building. Most of these failures are buildings next to buildings under construction. Almost always, the excavation for the new building had been carried several feet below the foundation of the other building.

Sometimes, as in the First National Bank, in New York, N.Y., an older building had been underpinned to what was thought to be a satisfactory level, but earth pressure against the underpinning had been enough to bring about failure. An occasional failure, such as the Meyers Department Store, occurs during alterations.

Often, minor failures—such as some settlement causing plaster cracks, or binding doors—are caused by construction operations in the building next door, even with the most careful conduct of the work. These minor failures frequently involve the parties doing the new construction in litigation.

Meyers Department Store—Albany, N.Y.

Although the failure described here dates back more than half a century, the lessons to be learned—and there are several—are just as applicable today as they were in 1905, when it occurred.

To begin with, the building [66] was put together from a group of old residences of uncertain age, between party walls, altered from time to time with no record of the various alterations. Much of the essential construction had been covered with plaster, thus misleading the casual observer into the belief that he was making alterations in one homogeneous building, instead of a heterogeneous group. The most important alteration had been made in 1881 when the party wall down the center of the 56 × 200-ft group of buildings was removed and replaced by a row of cast-iron beams and columns. Working backward, it is evident that the original buildings must have dated back to 1850 or earlier. Moreover, the variation in this construction indicates that even this 1881 project may have been carried on in different stages. Despite its faults, the superstructure was not the primary cause of the failure, although it may have been a contributing factor—particularly in the extent of the collapse.

The building, which was composed of units varying from three to six stories high, was a full block in depth. Its frontage was apparently two residences wide on each street, each roughly 28 ft wide. To get a straight line of columns through the store, the load had been picked up eccentrically on some of the columns. That is to say, joists were carried into the original brick wall; the upper part of the brick wall was left in place and picked up by beams or lintels that were off center of the wall, thus providing bending or lateral thrust in the columns.

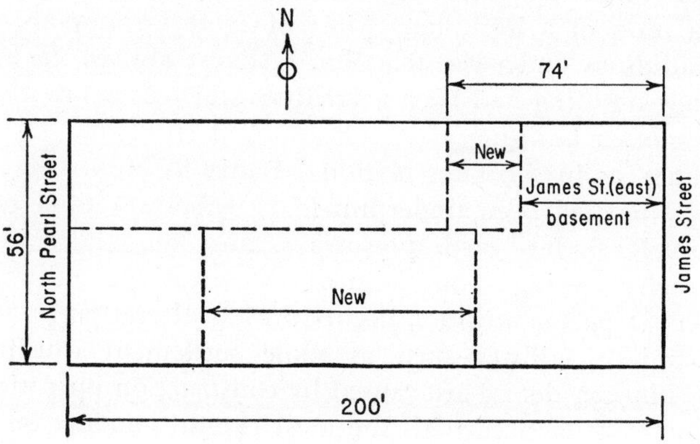

FIG. 6.1. Plot plan of Meyers Department Store

The original land was a steeply sloping site, so that the James Street (east end) level was almost two stories lower than the North Pearl Street (west) end. The east end of the store originally had a basement level that was a full story lower then the west end basement. The owners decided to run a subbasement for half the width of the store—an extension of the east basement almost through to the west end (Fig. 6.1 and 6.2).

The sequence of operations, some of which could be conducted simultaneously, involved: (1) underpinning about 70 ft of the existing south wall and a short portion of the north wall; (2) building new retaining walls along the west end of the subbasement extensions, and removing the existing west walls of this area; (3) excavating for the underpinning of the cast-iron columns along the center line of the building, and underpinning the construction to permit the extension of these columns; (4) removing the existing column footings and constructing new footings with 24-in. square brick piers to pick up the existing columns and form new supports from subbasement to basement floor; and (5) completing the excavation and placing a new retaining wall between the columns to form a new north wall of the enlarged subbasement.

Any underpinning operation is hazardous, particularly when it involves an area which is in use throughout the construction operations. The operations described in the sequence above are illustrated in Fig.

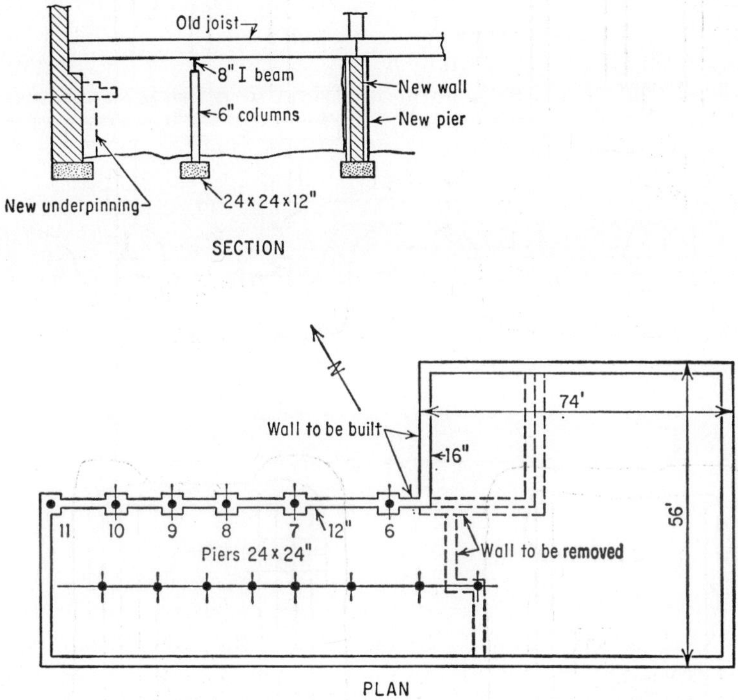

FIG. 6.2. Subbasement and extension of Meyers Department Store

6.3, and the situation just before the failure is shown in Fig. 6.4 and 6.5. The work was on a cost-plus basis, and therefore, the contractor had no incentive to cut corners in any way. Despite the type of contract, however, neither sheet piling nor shoring was used to contain the moist, clay subsoil under the column footings in question, the point at which failure began. The architect's drawing, which defined the work to be done, showed the plan of the reconstructed subbasement, two transverse sections, and various sketches of column details together with all necessary dimensions, sizes of walls, piers, and other details of the finished construction. The drawing did not, however, specify the manner in which the work was to be accomplished.

At the time of the collapse, column 7, as shown on the plans, had been properly needled and was fully supported on shores, but had not yet been underpinned. Work was underway on columns 8, 9, and 10. Excavation had been completed to a depth of 5 ft

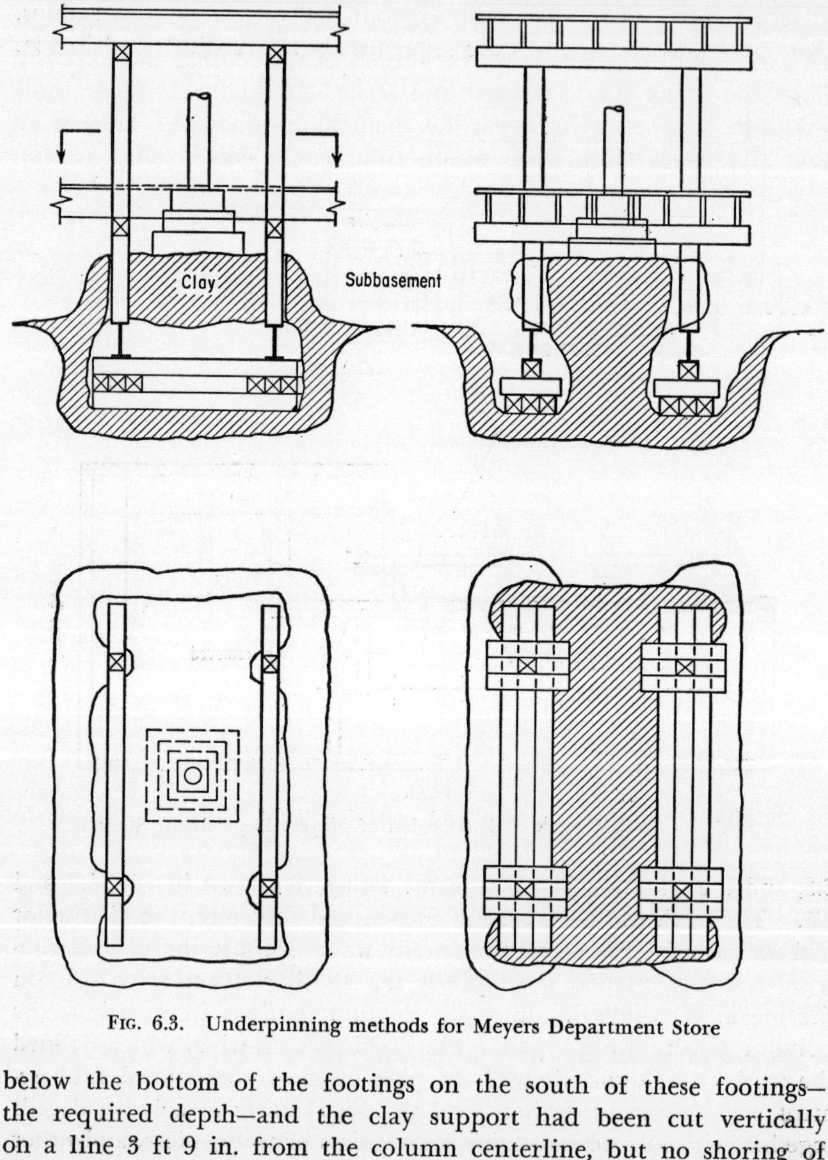

FIG. 6.3. Underpinning methods for Meyers Department Store

below the bottom of the footings on the south of these footings—the required depth—and the clay support had been cut vertically on a line 3 ft 9 in. from the column centerline, but no shoring of the floors and columns above had been placed. On the north side of column 8, the excavation had been carried to a depth of about 3 ft below the bottom of the footings, but the vertical face of the clay was about 8½ ft from the column centerline. The earth to the east and west of column 8 had also been removed as indicated in Fig. 6.4.

On the morning of August 8, when workmen started work, it was noticed that some of the earth under column 8 had sloughed off back to a point where part of the bottom of the footing was exposed. Then the upper floors cracked and settled gradually, but, for some unknown reason, the store was not immediately vacated. Within an hour after the first discovery of any trouble, the entire center section

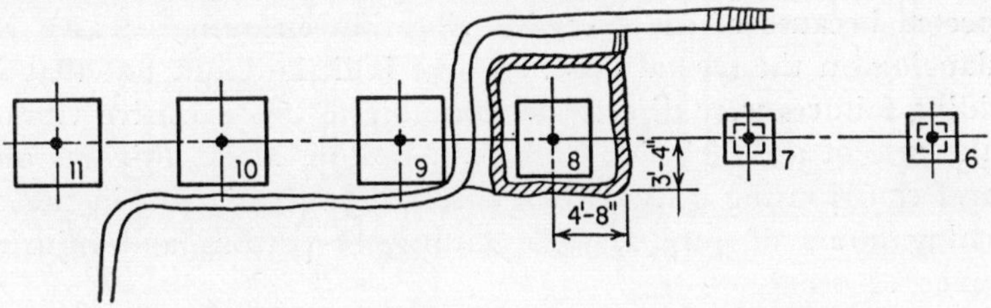

Fig. 6.4. Columns before Meyers Department Store failure—plan

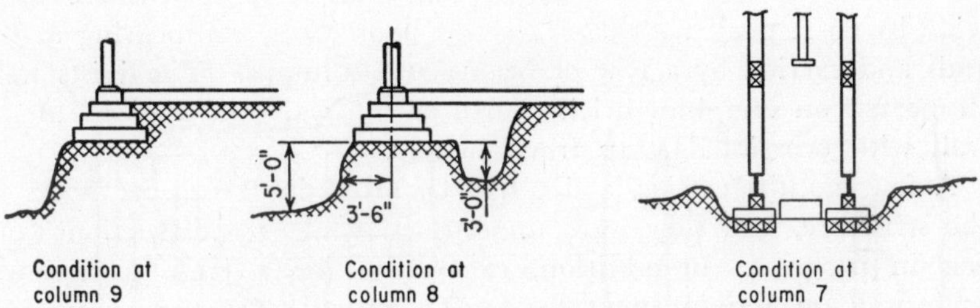

Fig. 6.5. Columns before Meyers Department Store failure—section

of the floor collapsed, beginning with column 8 and followed by columns 9 and 10. Thirteen of the store employees died but the workmen in the subbasement had time to escape to the east end of the building, and none of them were killed. In addition, a number of other persons were injured.

The mayor's special commission appointed to investigate the accident, reported that

> . . . the collapse was caused by the lateral crushing out of the bank of earth under one or more of columns 8, 9, and 10. It is our judgment that the conditions existing at the building and the character of the work in progress demanded that extraordinary precautions be taken to insure safety. If such precautions had been taken, the disaster could have been averted.

The report of the coroner's commission reported similar findings and the coroner's report went on to recommend that to guard against similar accidents in the future, the City of Albany establish a building department with competent superintendents and inspectors "to pass on plans and specifications submitted for their approval, and to see that any work proposed is carried out in a safe and workmanlike manner."

Pickwick Club—Boston, Mass.

Those who practiced architecture, engineering, or construction during the twenties when people were dancing the Charleston will recall the anxious phone calls of owners who wanted their buildings inspected because of the excessive vibration encountered as a result of dancing on the second floor. There is little doubt but that some building failures were directly attributable to this excessive vibration. In the case of the Pickwick Club [67] in Boston, Mass., this was one of several contributing causes when the building collapsed in the early morning hours of July 4, 1925, killing 44 persons and injuring a number of others.

The Pickwick Club, a night club of the Prohibition era, occupied the second floor of an old 5-story brick building, a structure 45 ft wide by 65 ft deep with 3×12 wood floor joists, wallbearing at the ends and carried by a row of beams and columns. The joists were supported on one 4-in. brick course of the wall and tied in to the wall with occasional strap iron anchors.

A few months earlier, a fire had occurred in the upper floors of the structure, and they were unoccupied while reconstruction work was in progress. In addition, excavation for a large garage was underway on the lot next door. To provide for the excavation, the masonry wall of the Pickwick Club building had been underpinned with seven concrete piers, 2 ft thick and 4 ft wide, and about 7 ft high.

The excavation had been carried down approximately to the level of the bottom of the piers. Apparently the wall, which was supported by the underpinning piers, had an earth pressure behind it to a height of 14 ft—the level of the basement floor. Thus the piers were furnishing vertical support for the wall but there was no provision for the lateral pressure against the wall. When the failure took place, the piers apparently tipped over, permitting the whole side wall to fall, thus collapsing the entire building.

Rua Rosario Building—Rio de Janeiro, Brazil

When engineers disagree on the safety of a structure, every precaution should be taken. Despite the fact that a city engineer had inspected the building only 30 minutes before and called it safe, the Rua Rosario Building,[68] an eleven-story building in downtown Rio de Janeiro, Brazil, collapsed on February 19, 1957, killing nine persons.

The 12-year-old structure was in perfectly sound condition, supported on concrete piers as foundations. Excavation for a new building was going on next door when the building suddenly began to develop serious cracks in both floors and walls. Despite the fact that the tenants of a bank, which occupied the lower three floors of the building, became alarmed and moved out—advising other tenants to do the same—the city inspectors insisted that the building was safe. It was determined later that the excavation for the new structure next door had removed the lateral support of the concrete piers, thus causing the collapse.

Lonsdale power plant—Lonsdale, R.I.

The economic loss to a small community resulting from the temporary closing down of a plant for reconstruction may be far greater than the actual loss of the work. The collapse of a power plant and a mill [69] of the Lonsdale Co. at Lonsdale, R.I., September 9, 1912, involved a property loss of $75,000, but it idled 600 employees for an extended period of time.

The failure was caused by the hydraulic undermining of the foundations of the water wheel, which had supplied the power for the plant for 30 to 40 years. The supports had undergone repairs only two weeks before the failure and it was apparently in sound condition.

First National Bank—New York, N.Y.

When settlement occurs under a building in the Wall Street section of New York, N.Y., the damages involved financially are commensurate with the main business use of the area. This was demonstrated on January 30, 1934, when the court awarded $237,000 damages to the First National Bank of New York for damages caused by excavation for the Bankers Trust Company Building,[70] next to the First National Bank.

The older building, situated at the corner of Broadway and Wall Street and known as 2–6 Wall Street, was carried on spread footings below the normal basement level. The building had settled somewhat twice before, first during excavation for the Stock Exchange Building when one of the pneumatic caissons for the new building blew in, releasing earth pressure, and again when the Irving Trust Company Building was built across the street.

In the construction of the Bankers Trust Company Building, eight underpinning caissons were carried down to rock under Number 6 Wall Street, the east section of the First National Bank Building, then sheet piles were driven, and excavation for the new building was carried on within the sheet pile coffer dam.

In spite of the precautions taken, a flow of subsoil from under the First National Bank Building caused settlement and cracks so extensive that the New York City Building Department ordered the building condemned and vacated. This was the cause of the lawsuit and the reason for the large award.

Department store wall—Topeka, Kan.

Normally an 8- to 10-ft berm—an earth embankment left in place to provide lateral support—should be enough to maintain a wall against lateral displacement, except under some very unusual circumstances. In the case of the failure of the rear wall of a department store[71] in Topeka, Kan., November 6, 1920, the unusual circumstances were: a weak spot in the berm at the center of the wall, caused by an old filled-in well; and the leakage of drain lines from the existing store near the same point. Both of these flaws were under the pier that supported the girders. They in turn carried the floor and roof joists, which ran parallel with the back wall.

The excavation was for the basement for an extension to the store. No indications of failure were evident until a few hours before the collapse when a night watchman heard cracking noises. The indications increased, and failure occurred about 8 A.M., when the back wall slid outward into the excavation. The girders and joists sagged but did not fall with the wall. Because of the time and the amount of warning, no casualties occurred, but large property damage was sustained.

Schmidt Building—Buffalo, N.Y.

On Saturday morning, September 20, 1924, a portion of the north wall of an old building occupied by Schmidt's Restaurant collapsed while excavation operations were in progress next door for an addition to the Lafayette Hotel. The existing building on the site had been removed, the excavation carried down to the necessary depth, and a series of shores placed against the existing old wall. The foundations of the old building extended to a depth below the new excavation.

The subsoil was a medium clay, and the new foundation was designed as a mat footing to carry the seven-story annex to the hotel. It was uncertain when the old building had been built, but part of the side wall indicated that it had originally been a two-story building, rebuilt at some later date into a high single-story structure.

In rebuilding, the roof structure had been carried independently of the wall on wood beams on steel channels, apparently supported either on a wood sill or on the stone foundation wall or by some other method not visible after the collapse.

About three-quarters of the length of the wall collapsed. A large crack opened up in the front wall through the first windows from the corner of the building; the upper part of the wall at this corner was in dangerous condition. The front top shore pushed its way partly through the wall, and several of the lower shores had a decided buckle in them. Beginning just back of the front corner of the building, the side wall collapsed for about three quarters of the length of the building. The wall pushed out into the excavated area about 6 to 8 in. at the base, and settled about 6 to 8 in., bulging the clay outside the wall.

Strangely, the second floor and roof did not collapse but hung suspended in spite of the wall failure, probably supported to some extent by the furring members inside the wall. Although a number of people were in the first-floor restaurant at the time of the collapse, no one was injured.

Bank excavation—Wheaton, Ill.

It does not seem possible that an experienced contractor would take the chances indicated or that a building inspector, architect, or anyone familiar with the hazards involved would permit the sequence of events that took place in the excavation for a new bank [72] in Wheaton, Ill., in August 1926.

The contractor engaged in the work had excavated 2 to 3 ft below the footing of the stone basement wall of an adjacent building and cut off the projecting stone footing preparatory to pouring his basement wall. No shoring or underpinning of the adjacent wall had been provided and the footing trench extended the full length of the old wall.

The material was a stiff, blue clay. The clay slipped out into the trench, dropping the end of the two-story brick-and-frame building more than 1 ft. The stone basement wall dropped almost vertically, but the brick end and front wall above were cracked and distorted; parts of the brickwork fell, and the end wall leaned dangerously over the pit. The building consisted of three bays, and the south bay had to be rebuilt. It did not fail, but wreckers knocked down the brick wall and then shored up the interior.

Factory building—Richmond, Va.

Buildings on a side hill can present unforeseen problems. A four-story mill building,[73] in Richmond, Va., had been built on a side hill; the older building on the downhill side next door to it had a basement that went down 10 ft below the foundations of this newer mill building, which had been built about 1912 (Fig. 6.6).

At some time during 1919, the older building was removed, but the uphill wall was left in place from the old basement floor to the street level, to act as a retaining wall and hold the ground under the newer building.

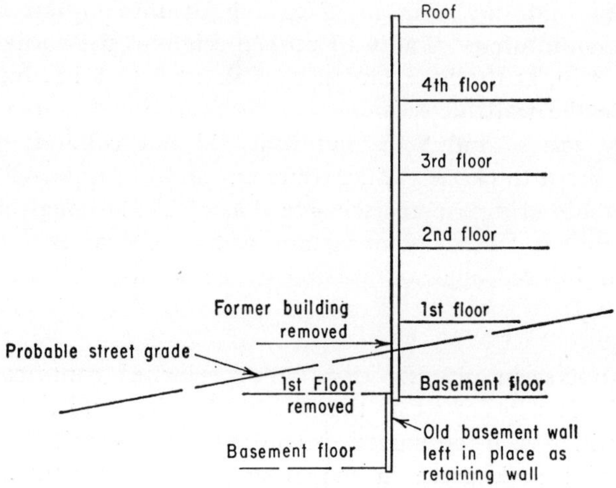

FIG. 6.6. Section of factory building—Richmond, Va.

The wall left standing had, of course, never been designed to take retaining-wall pressure, and certainly not the surcharge of a four-story building. On September 15, 1922, without any warning, the wall kicked out at the bottom and dropped the wall and adjacent floors of the four-story mill building. The failure happened during lunch hour and the few people in the building suffered only minor injuries.

6.3 Flow of Subsoil

An occasional failure may be due to the flow of subsoil from under a building, owing to the superimposed weight. Because of their frequent location on city waterfronts and their very heavy loading per square foot, grain elevators seem to be particularly susceptible to this type of failure.

City grain elevator—Portland, Ore.

Often the method of correction of a bad situation is of great interest to a technical man who may face the same problem at some time. The settlement of the Portland City Grain Elevator,[74] a part of the St. James Terminal at Portland, Ore., was arrested in 1919 by the use of "pinch piles"—piles driven to squeeze in the earth and prevent further outflow of the earth from under the elevator. The building consisted of three units: a track shed, an operating house, and a storage annex. The site was originally a flat about 10 ft above low water of the Willamette River approximately 7 miles north of downtown Portland. The grade was raised by means of hydraulic fill dredged from the river to bring the level to elevation + 32 ft, using low water as elevation 0. First the original grade was excavated to elevation + 4.5, then wood piles, 40 to 45 ft long, were driven 2½ ft apart in both directions over the entire site, a total of more than 4,300 piles. The piles were cut off at elevation + 5.0. On these piles a 3-ft-thick mat of concrete was poured, and pedestals were erected on the mat to carry the three parts to the elevator at their respective levels. The operating house and the storage annex pedestals were 16.2 ft long and carried a 15-in.-thick floor at elevation 23.8. The 16-ft space between the mat and the underside of the floor slab was then to be filled with hydraulic fill.

By the time the storage annex had reached two-thirds of its height and they began to pump in the hydraulic fill, the building began to settle. The settlement began in November 1918; concreting operations were suspended in January 1919, but the settlement continued until May. The settlement at this time was as follows for the four corners of the storage annex: the northwest corner, 1.96 ft; the southwest corner, 0.76 ft; the southeast corner, 1.40 ft; and the northwest corner, 1.15 feet. The settlement of the operating house ranged from 0.62 to 1.72 ft and the track shed from 0.21 to 0.37 ft. At the same time, it became evident that the entire area was settling; trees 300 to 500 ft from the building on which bench marks had been established had settled as much as 1 ft under the weight of the hydraulic fill.

The piles had been designed to carry 25 tons each, but the load on the track shed at this time was found to be only 15 tons per pile.

A board of consultants was appointed; they submitted a majority and minority report in April 1919. The disagreement in the two reports was not in the determination of the cause of the subsidence, but in the recommendations for arresting it.

The original exploration had been made by driving test piles over the site—a total of 14, ranging in length from 49 to 101 ft—to a maximum penetration of elevation −70. In the opinion of the board, however, the driving penetration—even at this depth—did not warrant the assumption of a safe load of 25 tons per pile, and they were unable to understand why piles 40 to 45 ft long could have been given this credit. Apparently no borings had been made to learn how deep the soft material extended before good bearing could be obtained. One pile that had been driven was test loaded to 40 tons, which was 1.6 times the design load. Also, a small group of piles supporting the dock were loaded to 60 to 70 tons per pile

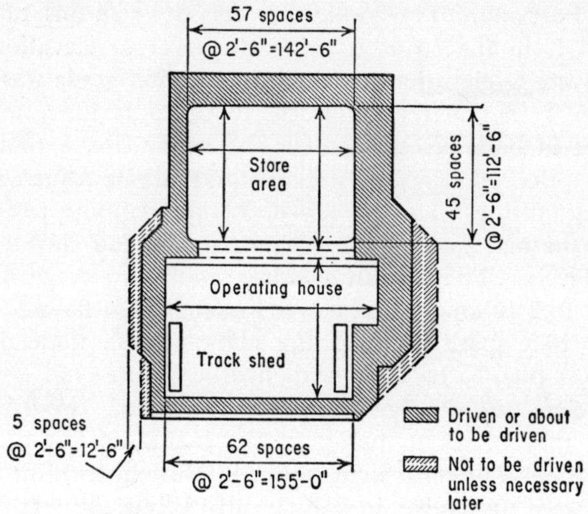

FIG. 6.7. Plan of grain elevator showing location of pinch piles

with no indication of subsidence. It was evident that the load on these test piles had been spread into the surrounding semifluid material by skin friction, and thus the load was dissipated. However, with a large number of piles closely spaced, this semifluid sand or silt would be greatly overloaded if all the piles were loaded to anything like 25 tons each. Over the total area in which the piles were driven, this would have amounted to 3.8 tone per sq ft on a material which probably would not have carried more than 1 to 1½ tons per sq ft. In effect, the piles were actually spreading the load into a material lower down, which was not capable of carrying as much per square foot as would the ground near the surface.

The majority report recommended that the outflow of the subsoil from beneath the building be stopped by surrounding the site with rows of closely spaced piles 15 to 20 ft deeper than the original piles in the outer rows, and crowding and solidifying the soil by starting driving at a distance of 25 ft out from the building, and driving successive rows closer to the building, thus "pinching" in the material and solidifying it. In effect, the operation would be similar to sheet piling the site to contain the soil. In addition to this, the report suggested that hydraulic fill be removed from the mat of the buildings to reduce the load (Fig. 6.7 and 6.8).

The alternative plan, which was the only plan concurred with by the minority report, called for placing concrete cylinder caissons as underpinning to a depth of about −160 ft, a total of 165 ft long, to a hard gravel underlying stratum.

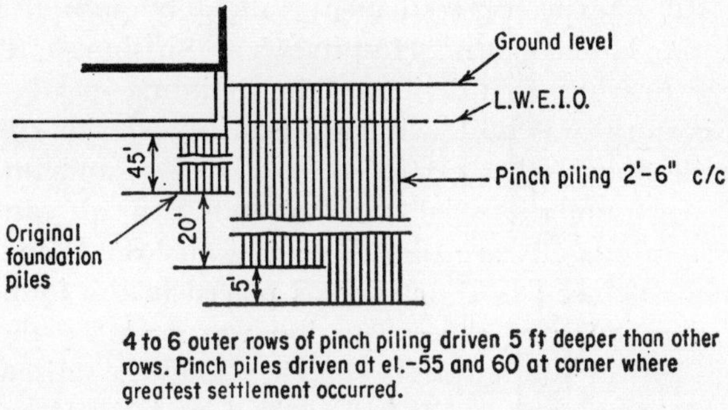

FIG. 6.8. Elevation of grain elevator showing pinch piles

The pinch pile method was chosen, and the longer piles were driven first around the northeast corner of the storage annex, where the settlement had been greatest. The subsidence was soon arrested here, and, as the driving came in closer to the building, it became much more difficult, indicating that the desired effect of consolidating the soil was being accomplished. As a result, the driving of the 2,600 pinch piles proceeded, and at the same time the construction work was resumed, apparently with no further trouble.

United Grain Growers Elevator—Fort William, Ont.

The case of the United Grain Growers Elevator [75] at Fort William, Ont., is peculiar because the building behaved properly for almost 30 years before giving any evidence of weakness.

A portion of this structure, about 100 × 385 ft, was made up of a 115-ft-high monolithic group of 56 bins, 4 bins wide and 14 bins long, extending outward from other buildings of the group into Lake Superior. The building had originally been built on fill contained by Wakefield sheet piling—a type of wood sheet piling built by spiking together three thicknesses of wood planking with the center member offset to form a tongue-and-groove piling.

In the summer of 1959, an inspection disclosed that some fill had been lost from the sheet piling enclosure, and about 500 cu yd of fill were replaced. About 24 hours before the failure occurred in October 1959, a slight lateral movement of the end of the elevator was noticed, but it was not considered dangerous at the time. With only enough warning to permit two workmen in the building to escape, the structure collapsed, dumping 2.5 million bushels of grain into the lake and causing an estimated damage of $4.5 million.

Homeopathic Hospital—Pittsburgh, Pa.

A rapid and unexpected settlement caused by subsoil movement occurred at the Homeopathic Hospital [76] at Pittsburgh, Pa., as the building was nearing completion. The building—partly four-story and partly six-story—was built with wallbearing exterior construction, and hollow-tile floor arches carried on steel beams and interior steel columns. The foundations of the columns are of concrete; the foundations of the walls are of rubble masonry on 12-in.-thick concrete footings. Before any concrete was placed in the foundations or footings, an inspection of the excavation was made by the architect and the president of the hospital, and they found a satisfactory hard brown clay. A test pit dug 4 ft below the foundation level disclosed the same material, and 1½-in. pipes driven with 16-lb mauls could only be driven to a depth of 4 ft.

After the roof was completed, the building suddenly began to show signs of distress. On June 2, 1908, a serious crack opened up in the wall, and by 7 P.M. that day the north wall showed a 4½ in. settlement. The settlement continued rapidly for the next two days and shores were placed to prevent a complete failure. By June 5, the settlement of the north wall had come to rest at 11¾ in., and the south wall had settled 2½ in. The interior columns showed no movement. At the same time, the north wall moved inward 9 in. and the south wall 1¾ in.

The footings were designed for a total load of 4 tons per sq ft and, at the time of the soil failure the wall footings were loaded to slightly more than two tons. The rainfall during May was above normal; rain fell on 24 of the 31 days. The downspouts had not been connected into the sewers and they spilled into the backfill between the foundation walls and the natural soil.

6.4 Insufficient Consideration of Subsoil

Settlement or sometimes even complete failure may occur because of neglect in finding out what the subsoil is, or, having this information, in not using it wisely. Proper test borings will reveal the nature of the subsoil and will avoid the danger of building an expensive building on a dump, as was done with the Mount Royal Pumping Station. Inspection on the job during construction operations would disclose the fact that the foundations were on clay instead of on shale, as in the factory building in Cleveland. A little investigation into what length piles had been used in adjacent buildings would have indicated that 69-ft piles would not suffice when surrounding buildings had required 86-ft piles, as in the Sao Luis Rei Building (Sec. 6.5).

Mount Royal Waterworks Pumping Station—Baltimore, Md.

The Mount Royal Waterworks Pumping Station [77] in Baltimore, Md., was designed in 1899 by a private architect employed by the waterworks and located on a site selected by the architect, over the objection of the water engineer. It was known that the site had, at one time, been a dumping ground for ashes and refuse of all sorts and that rock elevation varied from 19 to 47 ft below grade. The plans and specifications required circular piers of 5- and 10-ft diameter to be carried to sound bearing on solid rock. The architect supervised the entire job.

The building was completed in October 1899 and occupied in January 1900. About a year later trouble began to show up in various parts of the structure. The first indication was a cracked stone windowsill, followed by other masonry cracking. One crack in a masonry wall over an arch opened up to almost an inch. The cracking continued, and, when an addition was contemplated in 1907, the city building inspector was asked to make a thorough investigation. He found conditions that prompted him to have the foundations uncovered, and a serious condition was revealed. Foundation piers had settled as much as 6 in. away from the brick walls that they were supposed to support. When the excavation to the bottom of the piers was completed, some footings were found bearing on rock, or partially on rock, but others had as much as 3 ft of clay under the footing and over the rock.

During the time these piers were being placed, in fact when the piers had supposedly been excavated to rock, a severe rainstorm had caused a serious cave-in of the clay, and it is possible that this was the cause of the clay overlaying the rock. An extensive reconstruction operation was required to correct the situation.

Reinforced-concrete factory—Cleveland, Ohio

A newly built factory [78] in Cleveland, Ohio, began to show signs of serious trouble early in 1908, when the owner first started to use it. The design of the floor, where the trouble showed up, was interesting in that the bays were divided by beams into three spaces of about 8 ft each—the center space depressed 2 ft—so that the two side spaces formed platforms for the support of heavy tanks. Thus the main supporting girders at right angles to these beams were 5 ft deep at the ends, but only 3 ft deep across the center. When the tanks were being loaded, serious cracks developed at the edge of the depressed area, or rather at the edge of the beam that formed the side of the depressed area.

When an inspection was made the brick walls on the upper floors were found to be badly cracked, indicating a settlement of the interior columns. A review of the calculations showed that the foundations, which were supposed to be on shale, were loaded to 13 tons per sq ft. The building code permitted only 8 tons. Excavations were made to ascertain what the footing condition was, and they indicated that the interior column footings did not bear on shale, but on the clay overlying the shale. Although the girders cracked at their weakest point, it is probable that this failure was precipitated by a foundation failure. In any case, the building was repaired, strengthened and continued in use.

Equity Union Grain Elevator—Lincoln, Neb.

In 1954, shortly after it was completed and loaded, a section of the Equity Union Grain Elevator [79] at Lincoln, Neb.—a 5-million-bushel elevator—settled 8½ in. The section that settled was 283 ft long, only part of the length of the structure. Test borings showed that underground seepage from pockets and cavities in an underlying sandstone ledge caused the settlement. The condition was corrected and the building saved by mud jacking. The operation required the use of 41 carloads of a 6:1 mixture of sand and cement placed in a period of eight months.

6.5 Tipping of Buildings

One of the most interesting and spectacular types of foundation failure is the tipping failure. For a building to tip to the extent indicated by these articles, it must be quite rigid; otherwise some other type of failure would result.

There may be any one of several reasons for a tipping failure. For a heavily loaded grain elevator—these buildings seem to be particularly susceptible to this type of failure—the silos at one side of the building may be loaded to capacity with very little load in those on the opposite side. This causes an overload on part of the soil and perhaps an uplift on the opposite edge. A slight tipping resulting from this overloading of the soil will only increase the eccentricity and cause further tipping (Fig. 6.9 and 6.10). If the movement is not too rapid,

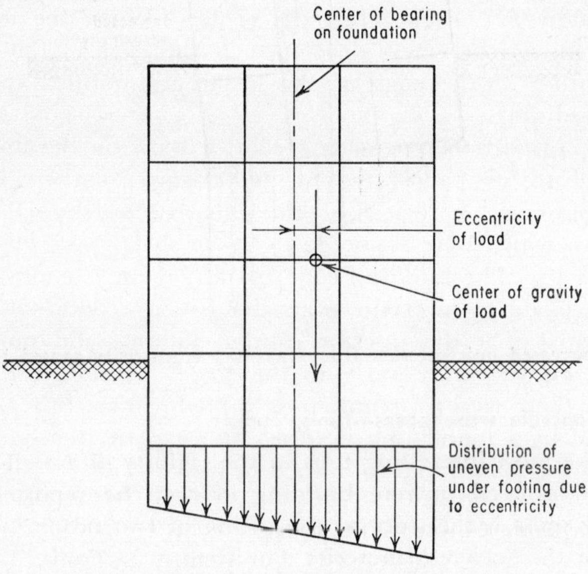

FIG. 6.9. Effect of eccentricity on soil pressure

this condition may be arrested by shifting the load to move the center of gravity to the opposite side of the building, as in the case below.

Another possible cause of tipping is a variation of the type or depth of subsoil under different sides of the building. If rock is 5 ft below the footing at one edge of the building and 30 ft below at the opposite edge, the compressibility of the overlying soil at the 30-ft depth is much greater than at the 5-ft depth and will permit the unequal settlement that can cause tipping.

Another possible cause of tipping is the release of pressure along one side of the building with no corresponding release on the opposite side, as might occur along the line of a subway construction or the installation of a deep intercepting sewer.

Some tipping of a building does not necessarily mean the death of the structure. Although an expensive corrective operation may be required, cases will show that by proper underpinning, a building may be saved.

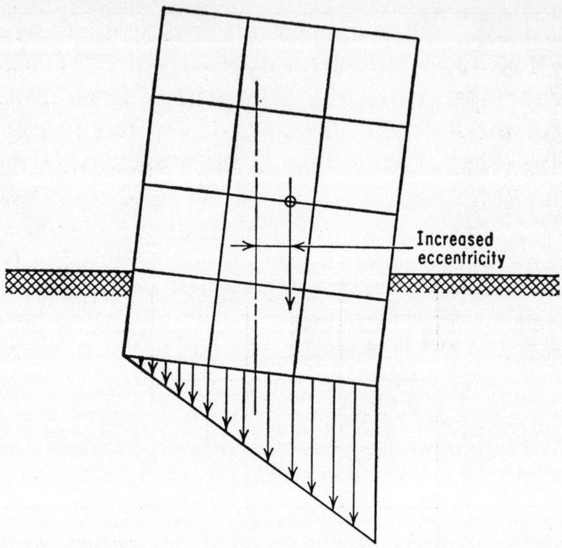

FIG. 6.10. Increased unit pressure from increased eccentricity caused by tipping

Reinforced-concrete warehouses—Tunis, Tunisia

One of the most interesting tests of the rigidity of a well-designed, well-built reinforced-concrete building ever to be reported in the engineering press is the story of the tilting of two newly built warehouses [80] for the Societe Minoteries Tunisiennes at Tunis, Tunisia, in 1906 (Fig. 6.11 and 6.12).

The buildings were two five-story warehouses—one 45 × 100 ft, the other 50 × 125 ft—parallel with each other and, between them but not connected, a one-story flour mill.

The construction men knew that the soft mud on which they were building was not good foundation material. A test well, 41 ft deep, was sunk without encountering any good material. Inverted arches, rather than piles were to be used to spread the load into a mat footing at a depth of 10 ft below grade. The middle building—the flour mill—was not included in the peculiar foundation design adopted and, because of its comparatively light weight, has given no trouble.

The foundation piers were brought up 32 in. A square framework of steel I-beams arranged in beam-and-girder formation was built upon the piers 30 in. above the ground, three beam spaces to a bay,

FIG. 6.11. Reinforced-concrete warehouse tilted by ground settlement—Tunis, Tunisia

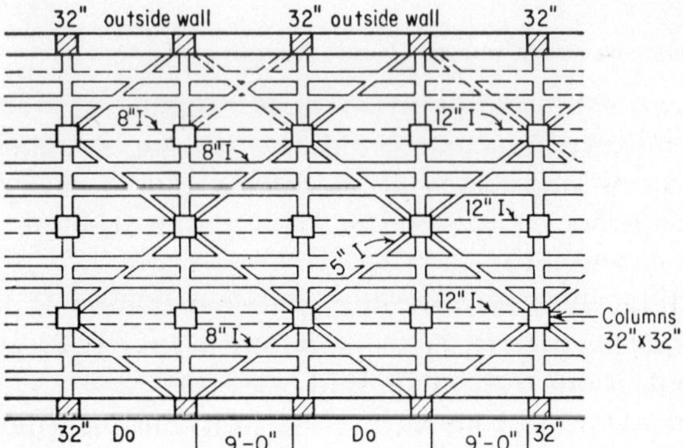

FIG. 6.12. Plan of first-floor steel framing of warehouse—Tunis

using 12-in. beams on column centers and 8-in. beams between, with cross-bracing across every column line with 5-in. I-beams. This network was formed with wood forms and concreted to construct a rigid base. Above this, the reinforced-concrete structure was built, using a 1:3 mix concrete throughout.

In June 1906, one of the two buildings began to tilt outward from the flour mill. It finally came to rest at an angle of about 25 deg from the vertical, but it showed no cracking or deformation. In an effort to return it to a vertical position, engineers loaded the basement and first floor on the high side with 2,000 tons each, a total of 4,000 tons of bagged sand and, at the same time, excavated under the high side to allow the soil to be easily displaced. In a few days, the warehouse was restored to its original position but not its original elevation.

In August, when the other building was being loaded for test purposes, it also began to tilt away from the center building. Because of the test, a transit was sighted on the building when it began to move under a uniformly distributed test load of 1,500 tons. The following readings were taken at intervals throughout the day—the readings being the overhang at the top compared with a given spot at the base—1 in. at 7 A.M., 3 ft at 2 P.M., 9 ft at 6 P.M., 9 ft 11 in. at 7 P.M., 14 ft at 9 P.M. and a total of 17 ft at midnight.

The building rotated around the center of its long axis, the level of the first floor at the high side being 9 ft above normal. The test load was transferred to the high side as in the other buillding, and the soil excavated as before. The building righted itself, but the two buildings were left 15 and 18 ft respectively below their normal levels.

Canadian Pacific Grain Elevator—Transcona, Man.

The Canadian Pacific Grain Elevator [81] at Transcona, Man., was one of the leaning-tower-of-Pisa types of failure; however, this was worth more to the owner upright than leaning. The operation of successfully righting it was carried out by the Foundation Co. of Canada.

The foundation—a concrete mat 77 × 195 ft—was placed in July 1912, after an extremely wet spring. At the time of placing, however, the material in which it was placed was dry and hard. Soil tests were made by loading the unit areas to 5 tons per sq ft with no apparent settlement. The storage bins were built on the concrete mat in 5 rows of 13 silos—a total of 65 bins with 48 diamond-shaped interstices, all attached to each other to make up a building approximately 75 × 193 ft. Each silo was 93 ft high and measured 14 ft 4 in. inside diameter with 6-in. reinforced-concrete walls. On top of this a cupola or head house for the conveyor system was erected. The bottom of the foundation was placed 10 ft below grade. The total capacity of the elevator was 1 million bushels, and under this load the total soil pressure from the live and dead load would be 6,620 psf. The working house was next to the elevator but was built on a separate foundation not connected to the elevator foundation.

After the failure, a thorough investigation of the subsoil was made down to bedrock and disclosed the following overlying strata: for 35 ft below the surface (25 ft below the footing) a stiff, impervious, blue clay, easily capable of carrying the superimposed load, then 5 to 8 ft of greasy, water-bearing, yellow clay mixed with gravel and disintegrated rock, then a layer of disintegrated limestone overlaying bedrock. The surface of the bedrock either sloped slightly or had shelves in it, which brought about a varying thickness of the soft clay.

The elevator was put in operation late in the fall of 1912 and showed no signs of distress until a year later, October 17, 1913. When the loading had reached about 850,000 bushels—85 per cent of its capacity, at which time the total soil pressure was 6,100 psf—the entire structure dropped vertically about 12 in. Prior to this, a slight settlement had been noted, but not enough to cause any alarm. With the dropping of the structure, the soil began to bulge up for a distance of 25 ft from the building. At about noon of the following day, the entire structure began to tip slowly about its long axis. On the side where the depth to rock was greatest and the layer of greasy clay deepest, a greater settlement took place. As shown in Fig. 6.9 and 6.10, the greater yielding of the soft clay on the deeper side started the tipping, and this eccentricity, in turn, brought about a still greater pressure, causing a continued flow of the subsoil. The tipping continued for 24 hours, at which time the elevator came to rest at a 27-deg angle from the vertical.

The entire building settled as a unit; the only damage to the structure was the loss of the conveyor cupola, which sheared off and fell to the ground. The structure was righted by means of underpinning with open caissons to rock.

Grain elevator—Fargo, N.D.

Shortly after midnight June 1, 1955, an 800,000-bushel grain elevator [82] at Fargo, N.D., tipped over and collapsed. The elevator, which had been completed in August 1954, consisted of 20 circular silos, 19 ft in diameter and 120 ft high, covering a 52 × 216-ft area. A head house extended the structure to a height of 198 ft. The foundation was a 30-in.-thick mat footing with interlocking sheet-steel piling driven to a depth of 18 ft around the perimeter.

Some settlement of the silos had taken place before the collapse, but it had not been serious and had been corrected by shifting the grain to correct the listing. No one saw the final failure. The owner at first expressed the belief that an explosion had caused the collapse, but investigation disclosed the fact that it was a subsoil shear failure, which permitted the entire foundation mat to tilt and the superstructure to tip. Frank C. Mirgain, Dean of Engineering at North Dakota State College at Fargo, investigated the failure for *Engineering News-Record* and reported that a movement of the soil to the south apparently permitted the north side of the foundation to settle, tilting the structure to the north and causing excessive longitudinal tensile stresses in the silo walls. The lower part of the silos broke into sections about 15 ft long, and the upper part was completely shattered by the fall. South of the structure, along its entire length, the ground was forced upward as much as 15 ft for a distance of about 60 ft from the structure by the shifting of the underlying soil.

Unity Building—Chicago, Ill.

Because of the nature of its subsoil, Chicago, Ill., has contributed much to our knowledge of foundation construction through the years. The Unity Building,[83] a sixteen-story office building in downtown Chicago built in 1892, started to settle toward two of its four sides within a few years after it was occupied. By 1900, it was leaning about 6 in. toward the south.

In 1910, a narrow gauge freight railway tunnel was constructed along the south side of the building, with its tracks 40 ft below street level. Very soon the settlement became more rapid, although very little cracking was evident. At the instigation of the building commission, new readings were taken and it was found that the leaning was increasing rapidly so that in November 1911, it was 27 in. out of plumb. The owner was ordered to make the building safe, and he engaged a consulting engineer to advise him on procedure. The consultant's examination indicated some very alarming conditions in the structure—conditions that had apparently existed since the building was first built. The structure consisted of cast-iron columns, steel beams, and tile arches. The cast-iron columns had not been faced, and rough casting faces butted against each other. Where holes on the flanges of columns above and below did not match, the bolts had simply been omitted, instead of the holes being reamed out. In almost no case was there complete bearing from one column section to another. Moreover, diagonal wind bracing had been omitted in floors below the third floor, where it was most needed.

All these faulty structural conditions required correction before underpinning the columns could begin. Underpinning, using open caissons, at the place where the worst settlement was taking place, held the building against further settlement. No attempt was made to bring the building back to plumb. It is difficult to understand how the building leaned so far without structural damage, under the circumstances.

Sao Luis Rei Building—Rio de Janeiro, Brazil

The failure of the Rua Rosario Building in 1957, described in a previous case, caused a storm of criticism of unsafe buildings and

FIG. 6.13. Tilting of eleven-story Sao Luis Rei Building

building practices in Rio de Janeiro, Brazil. One engineer stated that "if Rio were to get even the slightest earthquake or hurricane, half the city would tumble down like a house of cards." One of the magazines listed 20 buildings in Rio that had collapsed during the preceding ten years and predicted the failure of others.

Among the buildings then under construction and already in trouble, was the eleven-story reinforced-concrete Sao Luis Rei Building.[84] This building, 95 × 39 ft, was begun in 1955. The foundation was carried by 99 concrete piles, 69 ft long, driven in an old quagmire of clay and peat. Neighboring buildings were founded on piles up to 86 ft in length. When the construction had reached the tenth floor level, signs of mass settlement began to appear. A second shifting occurred in January 1957. In both instances, engineers inspected the building and agreed that nothing serious had occurred.

For another year the work progressed satisfactorily until January 1958, when the settlement began to assume serious proportions. An underpinning company was called in to correct the serious settling along the long side of the building at the rear on Friday, January 24. After a quick survey, the company began an extensive underpinning operation on Monday, January 27. In spite of round-the-clock operations to prevent further movement, the settlement was found to be increasing, and on Wednesday January 29 four adjacent three- and four-story apartment houses were evacuated. By Thursday morning the settlement rate along the rear edge was 4 mm per hour—roughly an inch in six hours (Fig. 6.13). At 7:25 that evening everyone was ordered away from the building and half an hour later the building toppled. Fortunately, there had been enough warning so that there were no casualties.

6.6 Destruction of Wood Piles

Some people mistakenly believe that wood piles kept continuously under water will rot out. As a matter of fact, just the opposite is true. Wood kept continuously immersed will last for centuries. The palaces of Venice, the old buildings of Amsterdam, and even a few of the Roman ruins in Great Britain will bear out this fact. Trouble will occur in a comparatively short time, however, if wood is exposed to alternate wetting and drying. For this reason the top of wood piles decay when the water table is lowered. Moreover, it can safely be assumed that the water table is being lowered almost universally. The fact that rainfall is being carried directly into sewers because of the construction of new buildings and the paving of greater areas means that less water is soaking into the soil. The installation of new and deeper sewer lines also brings about this gradual but sure lowering of the water table, and failures of the type described will, in all probability, increase as time goes on.

The fortunate features of this type of failure are that it can be corrected by underpinning, and that it is a slow gradual settlement which gives plenty of warning, and therefore does not cause a collapse.

Chicago River warehouse—Chicago, Ill.

An eight-story Chicago, Ill., warehouse,[85] built in 1872 and added to after a fire in 1881, began to settle in 1918. By the middle of 1919, the condition became serious. A crack in the west wall opened from the footing to the roof, and serious zigzag cracking occurred in the front wall. Window arches cracked, stone lintels were broken, and the bottom of one wall shoved out about 2½ in. To add to the seriousness of the condition, the entire building was heavily loaded with coffee and tea. The building had brick walls, cast-iron columns, wrought-iron beams, and wood floors.

The foundations were uncovered, and it was found that because of the lowering of the water table in the adjacent Chicago River, the top of the piles had almost completely decayed; the building was in imminent danger of collapse. Without unloading the floors, remedial action was immediately begun. To prevent the front wall from sliding out, it was immediately tied back in by means of ties to the third row of columns back, through anchor plates on the face of the wall, and these ties were pulled as taut as possible (Fig. 6.14).

After the wall was prevented from sliding out, the process of underpinning and bringing the building back to level was begun. Levels showed that the settlement ranged up to 7 in. By means of needle beams the wall load was transferred, and a 6-ft-long section of the footing was removed. The decayed piles were cut off to sound timber and a trapezoidal footing was constructed. When the new concrete had set, the space between it and the bottom of the wall was filled in with three rows of screw jacks, set so as to touch each other. After bringing these in to proper bearing, another 6-ft section was cut out and rebuilt in the same manner, and so on, until the entire length of the settlement was supported. Similarly, a nest of 24 jacks was set under the column. When the columns, the front wall, and the west wall were all on jacks, the operation of raising them simultaneously ⅛ in. at a time was begun. In this manner the floors and walls were all brought back to level, and the cracks were closed. The jacks were then removed, a few at a time, and the space they had occupied was filled with brick masonry.

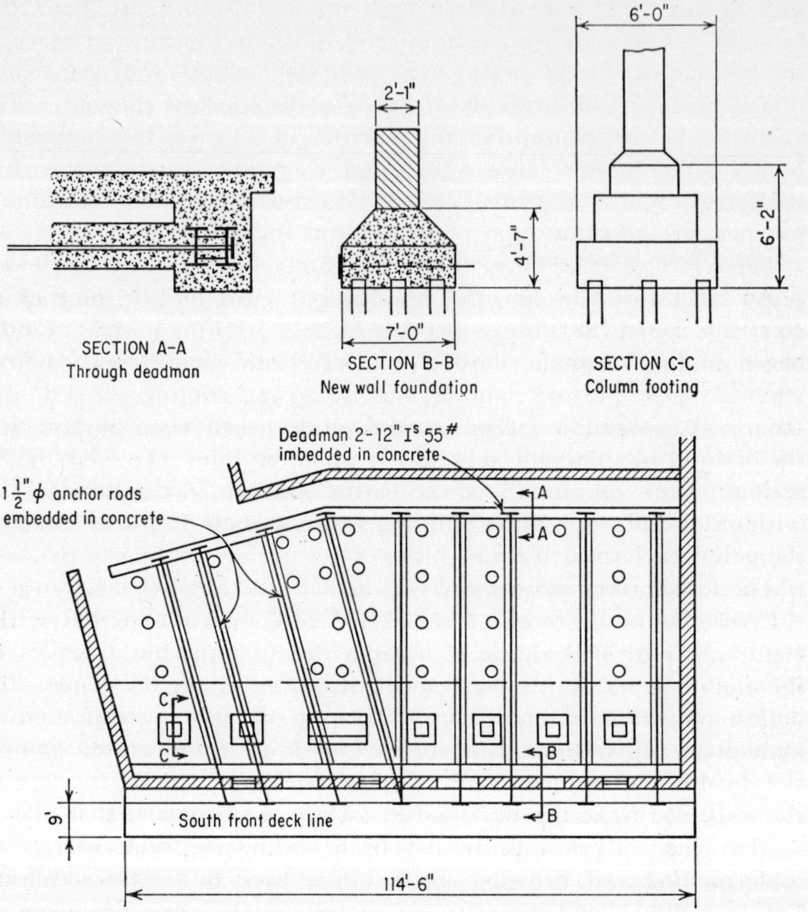

FIG. 6.14. Underpinning and anchoring for settled part of Chicago River warehouse

Millard Fillmore Hospital—Buffalo, N.Y.

Although not a complete failure, this case is included here because parts of the structure failed, and, had the condition been allowed to go on, a complete failure would probably have resulted. The author observed all phases of the work, and the case, therefore, is written up in more detail than it might otherwise have received.

In 1925 the Millard Fillmore Hospital in Buffalo, N.Y., added a four-story wing. Although the existing building had been built on good foundation material, a filled-in creek bed ran across the center of the site of the addition. Rock was located about 22 ft below the basement level, and it was decided to use concrete caissons to rock to carry the comparatively light column loads. The bids overran the estimate, and economical substitutions became necessary if the building was to be built. The contractor offered a $7,000 reduction if he could substitute wood piles for the concrete caissons. The engineer called the attention of the Board of Trustees to the possible hazard if the water table were lowered at some future date, but the Board decided that this possibility was so remote that the gamble was justified, and the wood piles were put in.

About twelve years later, the city of Buffalo put in a trunk sewer, just a block from the hospital, blasting out a channel in the rock to carry the sewer at the proper elevation. Within a year trouble began in the hospital. Partitions cracked and doors and windows jammed. An excavation made next to a wall footing showed that the top 12 in. of the wood piles had deteriorated to wood pulp so that the pile cap was acting as a spread footing, the piles themselves being useless. This condition did not exist uniformly throughout the building. Both ends of the building stayed at their original elevation; the settlement occurred through the center of the building in the area where the old creek had crossed the site.

Level readings were taken until 1945 when the differential settlement was more than 4 in.; it was decided to underpin the part of the building where the settlement was the worst. Although the building was not in immediate danger of collapse, a steam line or some other utility connection might have broken and caused panic.

Although the correction of resultant damage is not strictly within the scope of this book, the fact that others may be confronted with a similar problem prompts the author to include a brief description of the method used. It was suggested by Linton Hart of the Raymond Concrete Pile Company, and has served for 16 years in the area underpinned.

Thirteen-in.-diameter holes were shot drilled through the existing pile caps, and 10¾-in. pipe piles were driven through these holes. Shot drilling, which corrugates the sides of a hole, differs from core drilling, which gives a smooth-finished hole. The piles were driven in short lengths by a pile driving head suspended from an electric lift truck; a 5-ft length was driven, another length was welded on and driven, and so forth, until rock bearing was reached. The pipe was then cleaned out and filled to the bottom of the pile cap—also the top of the pipe—with concrete, and the pile was filled through the depth of the pile cap with concrete made with a mixture of cement and ironite compound to cause a slight volumetric increase in the top of the pile, thus picking up the corrugated surface of the pile cap.

New York Criminal Courts Building—New York, N.Y.

During the latter part of October 1909, the New York County Criminal Courts Building,[86] which had been built between 1890 and 1894, aroused New York citizens by its rapid and increasing settlement to the extent that on November 3, the New York Superintendent of Buildings ordered the building vacated.

It was a four-story building about 200 ft square with tile arch floors on steel beams, carried by stone and brick exterior and interior bearing walls. Up to 1801, it had been the site of the Collect Pond, which was filled in at that time. The west half of the building is on sandy subsoil, and the east half is on marsh or fill. Rock is about 100 ft below grade.

The foundation was built on wood piles, apparently of ample capacity under the original condition. Settlement defects began to appear shortly after the building was occupied and the cracking became more marked after 1900, when a subway was built west of the building. The western half of the building apparently remained stable, but the eastern side settled from 4 to 5 in. at its edge, opening cracks in the north and south walls. When a subway excavation along the east side of the building was begun in 1907, the condition became worse and new and larger cracks appeared, breaking an 8-in. water main within the building, causing plaster to fall, and general deterioration.

The subway floors were known to be lower than the top of the wood piles, and during the construction work, the piles were partially in dry soil. It is not known, however, to what extent deterioration of the building was caused by the lowering of the water table and consequent damage to the piles, or to settlement because of poorer subsoil on the side toward the filled-in pond.

City engineers, after further examination, recommended the installation of satisfactory shoring and regular weekly inspection for a year or more before underpinning was undertaken or defects in the building repaired. After three weeks of nonoccupancy, the building was again put into use.

Loft building foundation—St. Paul, Minn.

Mr. C. A. P. Turner, prominent Minneapolis consulting engineer, described a building settlement[87] which had come to his attention some time ago, where the failure of the wood piles occurred not from dry rot on top of the piles but rather as a result of the complete removal of the piles, through a distance of four or five feet, by wood borers.

The building had been built in 1886, on what had apparently been a swamp bottom of soft clay, with some fill. It was a four-story masonry structure, about 100 × 125 ft. The footings were in groups of nine pine piles to each footing; the piles were about 40 to 50 ft long. The pile groups were capped with dry stone foundations on which regular masonry piers were laid in natural cement mortar to the basement level, which was 8½ ft above.

In 1912, two additional stories were added and, at that time, it was noted that the building had settled from 2 to 3 in. in some places. This was attributed to a lowering of the natural water table through the installation of sewers and other construction through the years and the condition caused no alarm. No immediate change was noted when the two stories were added but, after a few years, when an additional 3 to 5 in. of settlement had occurred, an investigation was begun.

The investigation disclosed that the top of the piles immediately under the dry stone footing had decayed in the dry earth. Further down, however, the clay was still damp and plastic, and here, for a depth of 4 to 5 ft, only a round hole remained. Investigation revealed that the remaining portions of the piles below the tunnels made by the borers were intact. The beetles had probably deposited their eggs on the outside surface of the damp wood, probably 2 ft or more below the foundation, and, as the ground water level went lower, they had followed down—completely removing the sound wood from the hole.

The outcome of the discovery was a notice from the building department lowering the permissible live load on the building. Under this lower loading, it was estimated that the compressed clay would carry the load applied to the stone pile caps without any aid from the piles.

6.7 Subsidence

Town buildings—Elk Point, S.D.

Very serious subsidence conditions caused by lowering of ground water level imperiled the larger buildings [88] of the Missouri River town of Elk Point, S.D., in 1956. One of the most seriously affected buildings was the high school, where buckled walls forced the evacuation of the school and immediate shoring. Cracks of 4 to 8 in. wide opened between the walls and floor of one of the churches.

The citizens placed the blame on the construction of the Fort Randall Dam and the Gavins Point Dam, 75 and 40 miles up the river from Elk Point, stating that the water level in the river had been lower since these two dams were built. Army engineers, however, disputed this and blamed a long drought, as the rainfall had been below normal every month for more than two years.

6.8 Failure of Foundations

On rare occasions, the failure of a building is caused by the failure of the foundation itself, not from any external cause, but because the foundation was not strong enough to stand up under the load placed upon it.

American Ice Co.—Baltimore, Md.

Failures are seldom blamed on frozen concrete in the foundations, but this reason was given for the failure of a two-story brick building [89] under construction for the American Ice Co. in Baltimore, Md., December 14, 1911. Two workmen were seriously injured in the accident.

Other reports pointed, however, to improper foundation subsoil, caused by winter conditions. In either case, either the foundation failed because of delayed set of the concrete, or the subsoil failed under the partial load of construction. If the latter is the case, it is probable that failure certainly would have taken place sooner or later under full live load and dead load of the completed building.

Amsden Block—South Framingham, Mass.

About half of an almost completed building [90] failed in South Framingham, Mass., July 23, 1906, causing the death of twelve men. The three-and-one-half story Amsden Block was a building 75 × 120 ft, and was to have been one of the finest buildings in the town.

The columns were cast iron, the foundation and floors were concrete and the slabs were reinforced with wire mesh. The walls were hollow concrete blocks that had been cast on the site, and tests indicated that they were of excellent quality. The concrete was made up of one part of cement, three parts of clean but not well-graded sand taken from the excavation, and five or six parts of gravel brought to the site in sacks.

As far as could be learned, the collapse was a foundation failure. The slabs, despite the failure, were of good quality concrete. None of the cast-iron columns failed. Rumors spread that quicksand lay under the site, although none was uncovered at the level of the footing courses. The concrete foundations had a good, sound exterior, but the inside of them was soft enough to be scraped with a shovel. For some unknown reason, the core of the foundations had never set properly, permitting the cast-iron column bases to break off, so that under the dead load of the building the cast-iron shafts poked down into the soft interior concrete of the footings, fracturing the footings and precipitating the collapse. The details of the cast-iron base are not given in the magazine description of the failure, so it cannot be determined whether the actual cause may have been the underdesign of the cast-iron base.

Masonry pier—Boston, Mass.

Shoddy workmanship has caused many deaths throughout the years. Although the details of neither the construction nor the failure exist, the report by the building commissioner tells of the masonry pier [91] failure that killed three men in Boston, Mass., October 19, 1911.

The collapse was occasioned by the sinking of an iron column supporting the corner of the building and carrying on steel beams about 20 ft of wall on each of the two streets.

Upon uncovering the columns and the foundations under it, it was found that the column had sunk about 4 ft into the masonry pier built to support it. The column is of cast iron, 7 in. in diameter, with a shell 1 in. thick standing on a cast-iron plate 16 in. square and 1 in. thick. The column was filled with cement, and faced off square on the bottom. It was found intact, and in good condition. The iron plate upon which it stood was broken in five pieces—the central piece being carried by the column down through the foundations below.

Beneath the iron plate was a brick pier 2 ft 4 in. square and about 18 in. high, and beneath the brick pier a stone foundation was built, with a few square stones on the outer side toward the two streets and with small stones at the back against the old cellar walls, with many hollow places. The brickwork above the stone was laid up with little if any bond, and both stone and brickwork were laid in very poor mortar. The brickwork, when uncovered, showed moisture, and it seems apparent that the rains of the days preceding the collapse had softened and, in a measure, washed out the mortar in both the brickwork and the stone under it.

The collapse was occasioned by the iron column being forced down through the iron plate, the brickwork, and the stone foundation under the brick pier to a large stone near the bottom of the stonework.

The weight carried on the column was something over 40 tons. The column, brick pier, and stone foundation, properly built and in good mortar, would carry at least one half more weight than was imposed on them, and ignorance or carelessness must account for the bad workmanship.

The larger stone being on the face of the work and the brickwork being well finished with cement on the outside, the appearance of the work when finished did not suggest to the inspector the miserable construction in the interior of the pier. As the other masonry in the building seemed well done, there seemed no reason for suspecting bad work at this point. The pier was built between visits by the inspector, and was not seen by him until finished with the columns set upon it.

It is impossible for an inspector to see all the work in every building in the district when, as in the present instance, there are about 75 buildings and alterations in progress subject to his inspection and I find no reason for placing responsibility for this deplorable accident on him.

To guard against similar accidents in the future, it may be necessary to compel the employment by the owner of a qualified and licensed clerk-of-the-works who shall be responsible for the construction of the foundations, wall, floors, and all structural parts of a building.

Chicago Club—Chicago, Ill.

On June 17, 1928, a section of the Chicago Club [92] in Chicago, Ill., failed while alterations were in progress. The five-story building, with an area 80 × 100 ft, was built in 1886, with heavy load-bearing exterior walls and with flat tile-arch floors carried on steel beams and girders, which were in turn carried by cast-iron interior columns and wall piers spaced about 25 ft on centers in both directions. At the time of its construction, the building represented the most modern thinking in building design.

In July 1927, construction was started on a new ten-story annex, and the remodeling of the old building. The annex, a modern steel-frame-skeleton structure on caissons to rock, was completed and occupied in May 1928, and work on the remodeling of the old building was begun after it had been vacated.

At about 5 P.M. on Sunday, June 17, five floors of an area about 50 ft square, consisting of one cast-iron column and the bay each side of it in both directions collapsed. Strangely, the roof did not go down with the floors. Because the building had been vacated and the accident occurred during the weekend when no workmen were on the job, no injuries or deaths occurred.

As the wreckage was removed, a thorough study was made by a group of architects and engineers to determine the cause of the failure. The time of the failure—just about a month after the beginning of the remodeling operations—naturally pointed to some phase of the work as a contributing cause, but the investigation indicated that this was merely coincidental. When the removal of debris reached the first floor level, the column was found to be standing in its proper position in the basement, but the further removal of debris to the top of the footing disclosed the astonishing fact that this section of the column was actually the first to second floor column—the basement section had gone through the entire footing and was resting with its cap flush with the top of the footing. The column footing where failure had occurred is shown in elevation in Fig. 6.15.

It became evident, therefore, that the collapse was entirely a foundation failure. The soil for 30 ft below the footing is a soft, wet clay. Although the face of the masonry pyramid was in good condition, the material inside this sound facing was built up of comparatively small stones, laid up in lime mortar which had deteriorated during the 42 years since its construction. Much of the mortar had been washed away, and the remainder had very little strength. The cast-iron cap

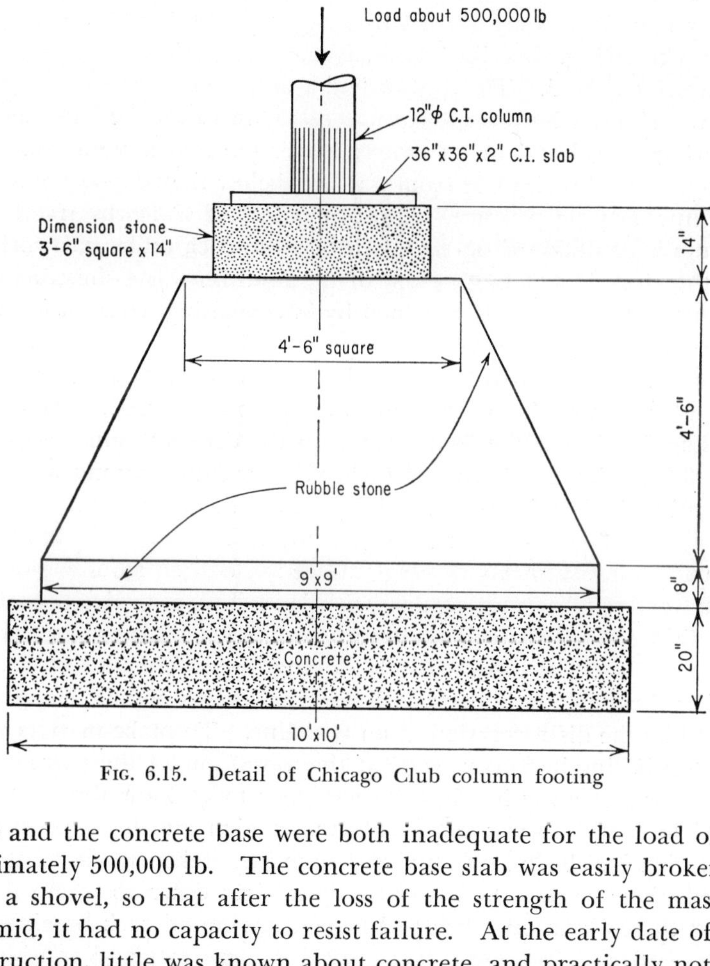

FIG. 6.15. Detail of Chicago Club column footing

plate and the concrete base were both inadequate for the load of approximately 500,000 lb. The concrete base slab was easily broken up with a shovel, so that after the loss of the strength of the masonry pyramid, it had no capacity to resist failure. At the early date of this construction, little was known about concrete, and practically nothing about reinforced concrete.

As for the cap plate, it is an interesting problem to compare this cast-iron plate with the modern steel billets required by good engineering practice today. A quick check indicated the need for a steel plate almost 4 in. thick instead of the cast-iron plate 2 in. thick.

A contributing cause to the collapse, undoubtedly, was a very heavy rainstorm, which had flooded the basement only a few hours before the accident. Further investigation disclosed the fact that a similar disintegrated condition existed in other footings, and as a result, the decision was made to remove and replace the entire building.

6.9 Improper Location of Footings

Very few failures are caused by improper location of footings; the Ireland Building, described later, is the only serious one the author has ever uncovered. The author had a sad experience many years ago that was similar, however. He was called in to do the structural engineering on a hotel in an eastern city. On a rectangular site, test borings showed rock to be from 2 to 8 ft below the proposed basement floor line, sloping practically uniformly toward a nearby river. The footings were designed on this basis to go to rock under each column, and the elevation of the bottom of the footings was assumed, for bid purposes, to be at a level obtained by interpolating between the elevations shown by the borings.

Excavation proceeded until one day a phone call came from the job. At the location of the first column footing to be poured, the workmen had dug 6 ft below the level shown on the drawings and had driven a rod 3 ft further but found no rock. By the following morning when the author had arrived on the job, the workmen had uncovered the rock along a 30- or 40-ft length of the wall. Instead of the rock showing a smooth, uniformly sloping surface as deduced from the borings, it was very rough and irregular, with jagged edges projecting up and soft, disintegrated material between hard layers. Instead of being in horizontal layers, it was in steeply sloping layers. Instead of the river flowing through the bottom of a syncline, it was actually in a valley formed by the disintegration of an anticline. To make matters worse, each test boring had been made at the top of one of the jagged edges, and a few feet away it might be necessary to go 20 ft deeper to find sound bearing. It was impossible to redesign any footing until the excavation had been made and rod soundings taken around it to find out whether it would be better to dig straight down to sound rock or to put the column on a bridge between two sound rock bearings.

The entire operation involved sizeable extra cost and considerable delay. Needless to say, the author was quite unpopular with the directors of the hotel corporation.

Ireland Building—New York, N.Y.

Although the failure of the Ireland Building [93] in New York, N.Y., occurred in August 1895 and is, therefore, ancient history, enough interesting facts occurred in connection with it to warrant including the story in any book on failures.

An eight-story, wallbearing building, it was typical of many of the buildings being built during that period. It was a 50 × 100-ft rectangle, with a row of four round cast-iron columns down the middle, thus dividing the area into bays of 20 × 25 ft. One line of girders composed of two 15-in. steel beams ran down the center from front to back of the building, and 10-in. steel beams, across the building at 5-ft inter-

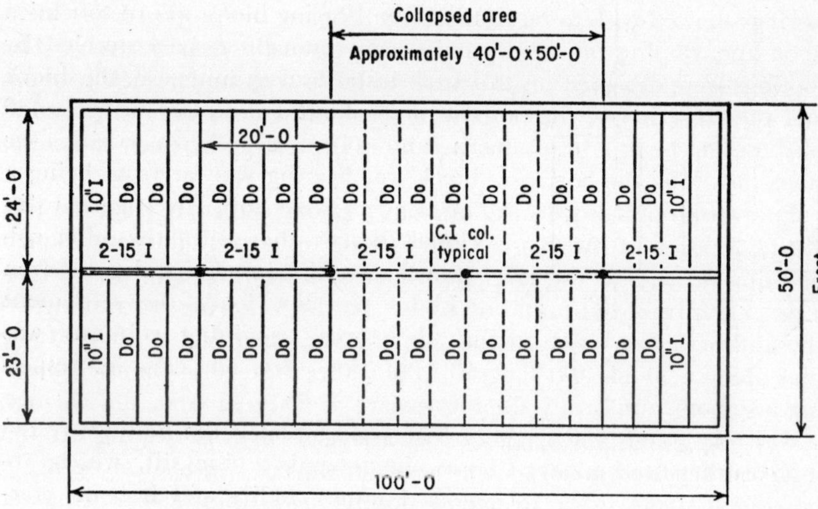

Fig. 6.16. Plan of Ireland Building showing framing

vals, framed to these girders—the outer end wall-bearing on the brick walls. These beams, in turn, supported the floor system—hollow-tile arches with concrete fill over the top of them to floor level, which was accepted at that time as standard fireproof construction. Figure 6.16 shows the plan of the building.

At the time of the failure, the entire structure was completed, and finishing operations—such as plastering—were under way. The area that failed, as shown on the drawing, was an area of 40 × 50 sq ft, all of which was dependent on one central column for support. This entire area, for the full width of the building, collapsed into the basement, causing the death of fifteen workmen and injury to others.

Several theories for the failure were immediately put forth: (1) the undermining of the foundations by an underground stream; (2) flaws in the cast-iron columns; (3) overloading of the floors with construction materials; (4) an improper foundation design; and several other reasons akin to the above.

Some weeks later, when the findings of the coroner's jury were released, it was learned that a foundation failure was at fault. The drawings had called for the columns to bear on a 3½-in.-thick cast-iron plate, 32 in. square, having a granite block 6 ft square and 12 in. thick, which in turn was to be carried on a concrete foundation 6 ft 9 in. square and 18 in. thick. Under these conditions the soil pressure would have been about 5,000 psf. The conditions actually uncovered showed that: (1) the concrete was actually only 12 in. thick instead of 18 in.; (2) the stone bearing block was of soft mica schist (soft enough to be crumbled by hand) instead of granite; (3) the block was only about 4 ft square instead of 6 ft; and (4) the soil on which the block was bearing was far from being a 5,000 psf capacity soil. Investigation disclosed the fact that all of the other columns had settled in varying degrees even without the live load; the settlement of the three remaining columns was 1$\frac{5}{16}$ in., 2$\frac{5}{16}$ in., and 2⅝ in. respectively.

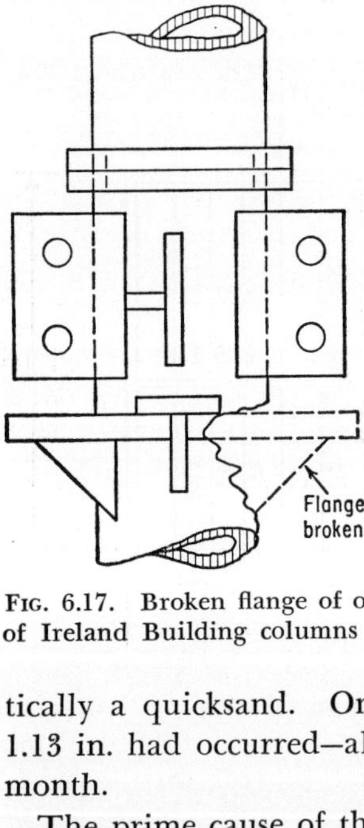

FIG. 6.17. Broken flange of one of Ireland Building columns

The soil on which the footings rested was, in general, a fine silt, which absorbed water readily and became practically a quicksand. On the day before the accident, a rainfall of 1.13 in. had occurred—almost one third of the normal rainfall for a month.

The prime cause of the failure was an old brick cistern under one corner of the footing that failed. This cistern extended under one side of the footing almost to the center. When the cistern was discovered, the cistern walls were torn down almost to the level of the underside of the footing; the cistern was filled with lean concrete; and the footing placed. Thus the bearing of the footing was partially on the concrete fill, partially on the brick wall of the cistern, partially on the backfill around the cistern—a material of questionable value—and probably partially on virgin soil. The concrete, stone, and cast-iron base had all split in both directions, and the column shaft had driven through all of them to a depth of almost 7 ft, thus pulling down all framing which had been attached to it. The pull on a beam attached to the adjacent column caused the failure of the column bracket (Fig. 6.17) because the resisting weight at the top of the beam connection was greater than the strength of the cast-iron bracket.

A pile of excessive length

Many years ago, while in conversation with a superintendent for the Raymond Concrete Pile Company, the author laughed over an incident that had happened on one of their jobs several years earlier. The superintendent had been driving piles for a building foundation in an area where salt had formerly been produced by pumping artificial brine from wells and extracting the salt from the brine, but where the industry had been abandoned.

For the benefit of those who are not familiar with the standard Raymond pile, it is a cast-in-place concrete pile in which a spirally reinforced sheet steel shell is placed on a retractable mandrel of tapered shape, and driven into the ground to become a permanent form into which the concrete is poured. If, because of idiosyncrasies of the material into which the pile is driven, it becomes necessary to lengthen a pile beyond the length of the shell that has been placed on the mandrel, the shell already driven is partly filled with concrete, and another shell is driven inside the first shell.

In this instance, the piles were practically a uniform length of 35 ft each, but one pile failed to act as it should. At 35 ft, it was still driving 10 in. to the blow. The superintendent partially filled the shell and drove again, still with no evidence of any bearing. When the pile was still going strong at 80 ft, he began to lose his temper, particularly when an old timer among the sidewalk superintendents kept chuckling at each successive failure. Finally, losing his temper, he asked, "What do you think is so funny about it?" The old timer replied, "Well, if you expect to put a pile there, you're going to drive from 900 to 1,200 ft before you fetch up. You're driving right on top of an old salt well."

6.10 Miscellaneous Foundation Failures

The cases described below do not fit into any of the previously discussed categories but are included because of their interesting aspects and the lessons to be learned.

Office building that slid downhill—Altena, Germany

An eight-story office building 44 ft wide and 174 ft long was being constructed on a steep side hill in Altena, Germany, in 1960 (Fig. 6.18).

The underlying strata were as indicated in the figure, a stratum of loose soil overlying a layer of quartzite, which, in turn, overlaid another loose stratum, which was carried by sound material underneath. The thickness of the various layers is not known but this is of less importance than the warning it carries and the clever way in which failure was arrested.

When the first five floors had been finished, the building began to move downhill—the loose top stratum sliding on the surface of the quartzite. The movement was indicated by cracks which appeared in the basement. The contractors worked fast to save their investment,

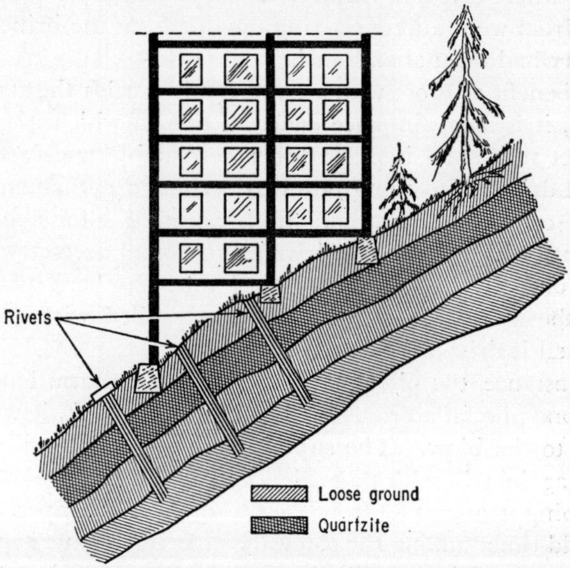

FIG. 6.18. Office building that slid downhill—Altena, Germany

drilling 150 2-in. holes, 20 ft deep, through the three top layers into the sound material below. In these holes, they set 1¾-in. steel bars, grouting them into the holes for their full depth. To these rods they built concrete walls to stop the loose material from sliding, thus containing the downward hill pressure resulting from the vertical weight of the building.

The steel rods were required to resist the diagonal component of the weight of the building in shear just as a rivet would resist shearing action, and the danger of collapse was abated.

Palace of Fine Arts—Mexico City, Mexico

Mexico City has, from its earliest days, produced plenty of problems for the foundation engineer. Typical of these problems—and perhaps the best known of them—was the settlement of the Palace of Fine Arts.[94] The city was built in a wide flat valley, about 7,400 ft above sea level. Although there are no major rivers in the area, the melting snows of the mountains and the rainfall became trapped in the valley, so that the water table was quite close to the surface of the

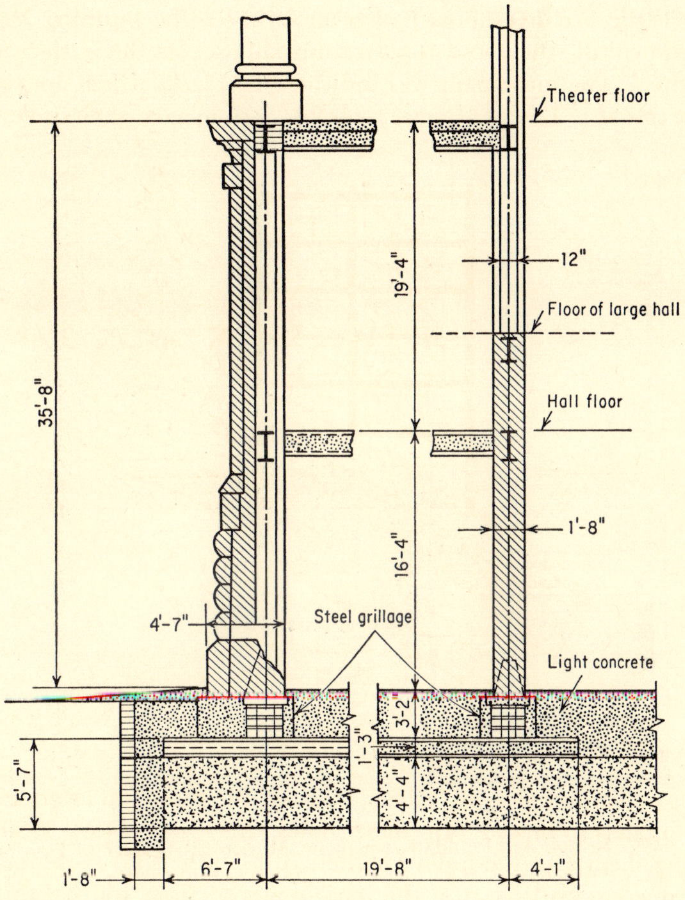

FIG. 6.19. Mexico City Palace of Fine Arts—section showing foundation mat

ground. In the seventeenth century, a drainage tunnel was built through the mountains to the east, and this tunnel, combined with the increase in water pumped from the ground as the population increased, lowered the water table and continues to lower it throughout the entire valley.

This valley was originally a lake with active volcanoes around it, and the entire area is filled with fine volcanic ash, which formed a spongy, compressible subsoil, capable of holding five to six times its weight of water. The elimination of the water has permitted a general settlement of the entire area as the volcanic ash has been compressed. When the general settlement is uniform, no damage is done, but, if the superimposed load over an area is great enough, the air spaces left by the dehydration are compressed and settlement of a more serious nature results.

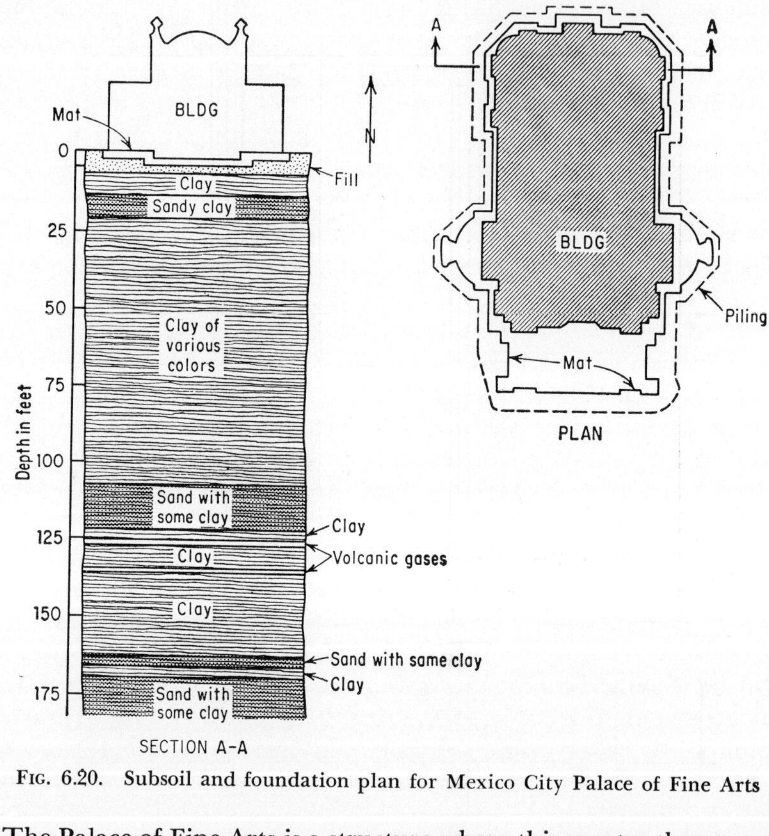

FIG. 6.20. Subsoil and foundation plan for Mexico City Palace of Fine Arts

The Palace of Fine Arts is a structure where this greater than normal settlement has presented problems. Begun in 1904, the construction, which gave trouble from the outset, was carried on spasmodically until its final completion in 1934. The building is of heavy masonry construction with a structural steel frame, and, because of the soil problems, is carried by a concrete mat 6 to 10 ft thick, with the load spread through the mat by means of 24-in. steel I-beams between the columns and 15-in. steel I-beams between them (Fig. 6.19).

The building is approximately 267 × 390 ft in outside dimensions, and the mat projects 5 ft beyond these lines except on one side where it extends 60 ft. The weight of the building is 57,500 tons, and the mat itself weighs 46,000 tons, so that the total load on the soil is about 1 ton per sq ft. When the original pour of concrete was made, a subsidence in the center of the mat was noted. The concrete from the bottom of the grillages to the top of the mat was then poured, using a lightweight concrete, and, with varying interruptions, the construction continued. A survey in 1907 showed that in addition to a general subsidence, the center had settled in a dished shape 1½ in. more than the rest of the building. By 1908, the part of the building already finished had settled 5½ ft, and two years later a diagonal crack appeared across the surface of the mat.

During the construction, attempts were made to stop the settlement by driving interlocking steel sheet piling, located as shown in Fig. 6.20, to prevent any flow of material from below the mat. Since the settlement is not the result of this type of mud wave, however, the sheet piling proved to be useless and the settlement continued at the same rate.

Because of the sequence of construction operations, a heavier load was applied at the north end of the building, and, during this period, the rate of settlement became greater at that end. In 1910 an effort to arrest the settlement was made by injecting cement grout into the subsoil at this end of the building. This method also failed to produce results, and recent investigations indicate that it may have done harm.

In 1950, the Western Foundation Corporation and Spencer, White and Prentis were engaged to make a survey and report (this article is based on a brief résumé of their lengthy report as printed in *Civil Engineering*).

During the intervening years, several skyscraper office buildings had been erected in this same area using long piles to a sand-bearing stratum 174 ft below grade, and these buildings—even the 44-story La Latino Americana Building next to the Palace of Fine Arts—have behaved satisfactorily.

The report therefore recommended the use of steel pipe piles jacked down to another apparently satisfactory layer 105 ft down, using 950 piles, each carrying approximately 100 tons. Apparently this work has not yet been undertaken and the settlement—now in excess of 10 ft—is continuing at the rate of 1½ in. per year.

7

Negligence and Ignorance

Negligence, carelessness, ignorance, greed! One or more of these factors was present in many of the failures cited throughout this book. Those included in this chapter are grouped here for various reasons. Perhaps negligence or greed—the most impressive factor—was the outstanding cause. Perhaps the failure was caused by a combination of concrete failure, masonry, steel, foundation—so many that it could not be put in any one of these chapters. The code of Hammurabi must have been an effective instrument to discourage negligence of this sort but, human nature being what it is, failures in which negligence, carelessness, ignorance, or greed are contributing factors will probably continue as long as men build.

License laws for architects and engineers were established to safeguard the public—public safety and property—and admittedly failures are fewer since license laws came into being. Whenever any law is proposed that will curb the right to do something, even though it may have a sound background, prolonged opposition occurs before the law can be passed and put into effect. This has been true of building codes, fire laws, license laws, and the like. Although a few states have passed laws licensing contractors, most have not, and, because of the number of individuals involved, such a law would probably meet with the greatest opposition of any proposed so far.

Negligence or carelessness often occurs because the men in charge do not properly supervise the work of others who may not realize the safety measures necessary—the failure to inspect the bracing of steel or of concrete formwork falls into this category. Sometimes the superintendent's nerves are so frazzled by a succession of problems, that his normal good judgment may not be functioning properly. In still other cases, as in some of those listed in this chapter, the cause is merely the desire to save every last nickel, even to the extent of taking a chance.

7.1 Multiple Causes

Knickerbocker Theater—Washington, D.C.

Probably the most publicized building failure of the twenties was the collapse of the Knickerbocker Theater[95] in Washington, D.C., on the evening of January 28, 1922, causing the death of 97 persons. The building had been built in 1916 and 1917, and wartime shortages had caused changes in the original structural plans, changes that had been approved by the Washington Building Department but which had apparently never been checked by the architect, nor by his engineer.

In 1921 the balcony was extended forward 18 in. to permit the addition of extra seats. Pairs of 4-in. channels were laid on the front platform of the existing balcony and anchor bolted into the riser of the first step—the new floor slab being carried by these channels. The steel structure of the balcony itself was not changed to provide for this addition, but there is no indication that this added load contributed to the failure in any way.

At the time of the collapse, the heaviest snowstorm since the building had been built was in progress and had deposited an average of 2 ft of snow on the roof. Because of the storm, only 400 to 500 people were in the audience—an unusually small attendance.

As shown in the framing plan of the roof, Fig. 7.1, the northwest or Columbia Road wall of the building was a curved wall, and the movement outward of the top of this wall was one of the chief contributing causes of the failure, as will be noted from extracts from the various reports which followed the collapse.

The two 12-in. channels forming the bottom chord of the main truss had only 8 in. of bearing on the west wall measured along the center of the chord (Fig. 7.2). They also overran the gusset plate 5½ in. Thus these two channels were practically independent of each other as far as bearing strength was concerned. This arrangement also placed the center of stress on the edge of the wall (which was flush with the edge of the bearing beam), producing an eccentricity of about 4 in. These two channels with their ends free, therefore, were subject not only to the end load of about 110,000 lb, but also to the bending stress caused by this eccentricity.

The web of the west channel at the end was still straight after the collapse, whereas that of the east end channel was rolled up. The

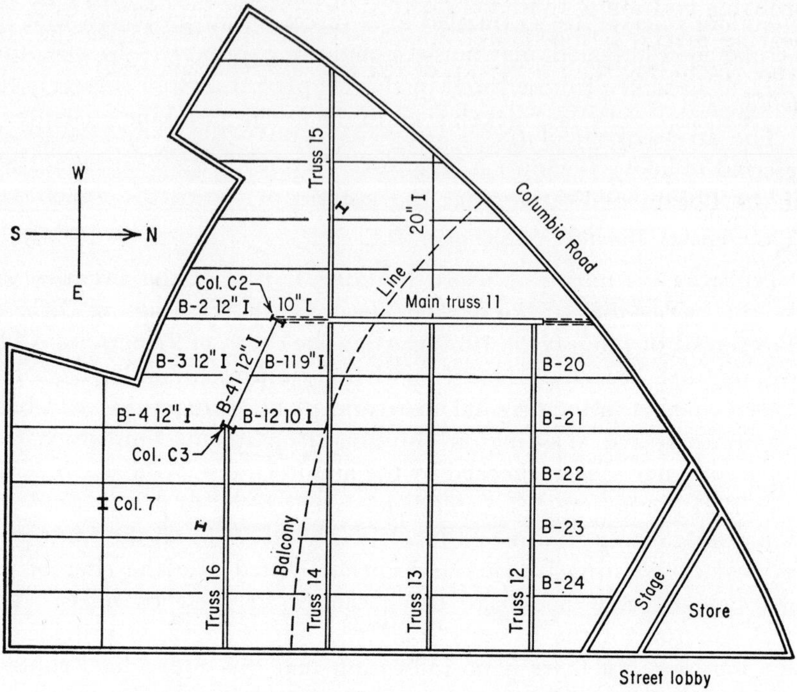

FIG. 7.1. Roof framing plan of Knickerbocker Theater

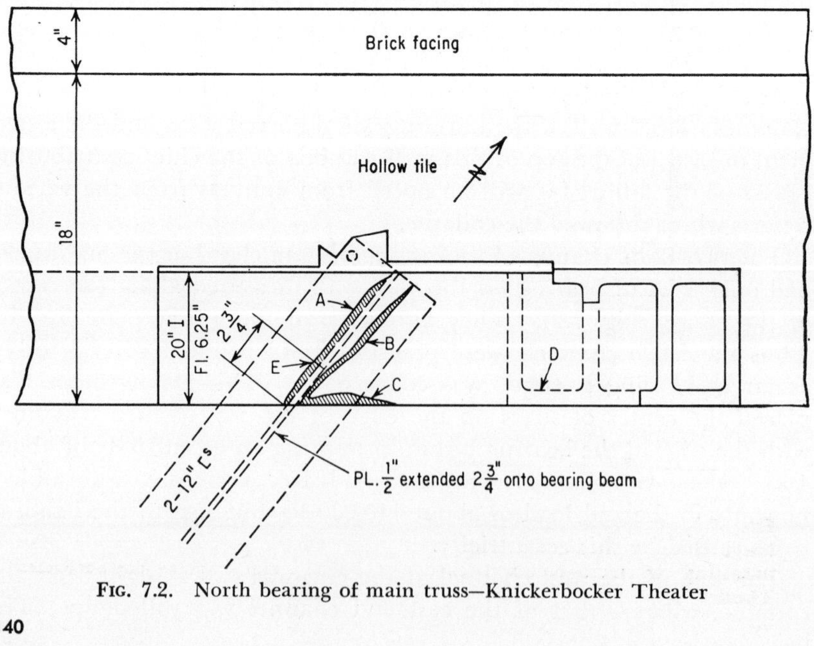

FIG. 7.2. North bearing of main truss—Knickerbocker Theater

probable position of these webs before the collapse (as indicated by the bearing marks) is shown in Fig. 7.3. Their condition after the collapse (including the rolling up of the east end channel web) is shown in Fig. 7.4.

The arrangement of the bearing that caused the 4-in. eccentricity referred to above is shown in Fig. 7.5. Marks on the bearing surface of the I-beam, indicating that the bottom of the east channel (not

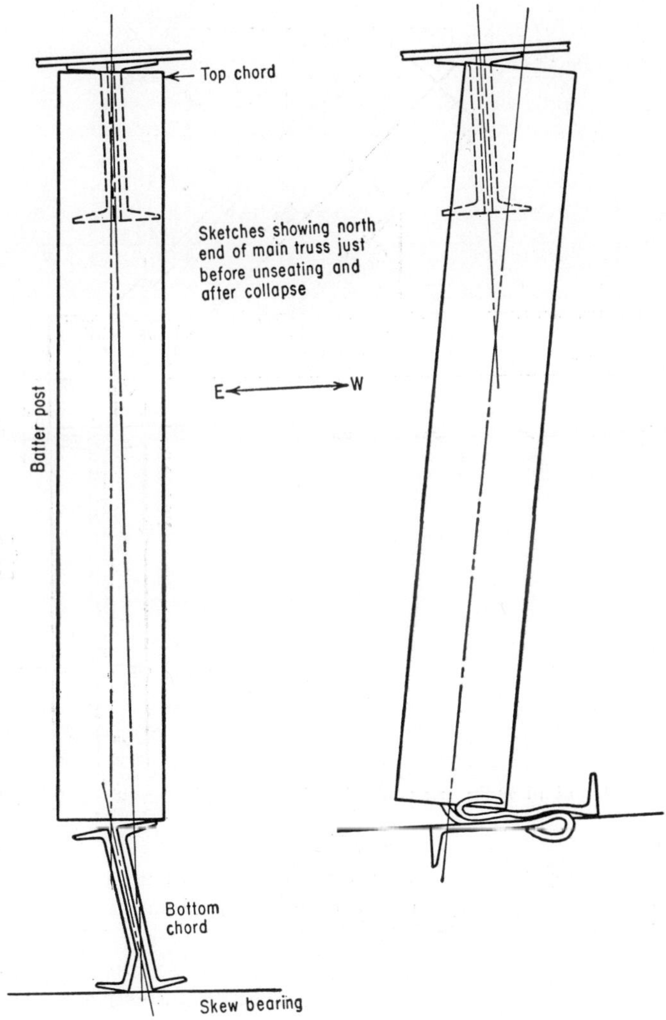

FIG. 7.3. Probable condition of lower chord immediately prior to unseating of truss—Knickerbocker Theater

FIG. 7.4. Approximate condition of batter post and chords after being cut apart—Knickerbocker Theater

shown in Fig. 7.5) had moved away from the west one before unseating, were plainly visible.

The following account of what happened was taken from the *American Architect–The Architectural Review,* for 1922. Details are shown in Fig. 7.2–7.5.

The roof fell bodily into the pit, carrying with it approximately the eastern two-thirds of the balcony. The column under the main truss, which did not

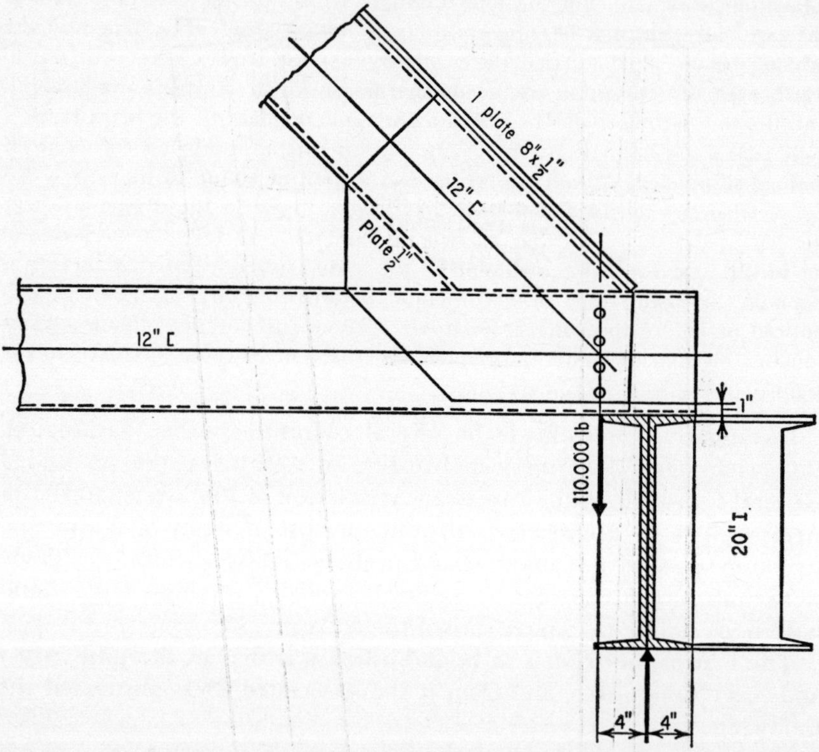

FIG. 7.5. West end of main truss and bearing at Columbia Road Wall—Knickerbocker Theater

extend to the roof, was pushed out from under the truss without being injured. This column rested on one of the balcony columns which did not fail and was joined to it by bolting through two spliced plates. The only column extending to the roof was buckled by the truss framing into it, and the load of this column fell upon the balcony, buckling the column directly beneath and one other column supporting the balcony.

The main truss was practically undistorted outside the north panel where the lower chord, which formed a bearing on the west wall, was buckled and twisted about 90 degrees. The gusset plate above this chord was buckled and torn while the batter post and top chords were thus thrown out of line about four inches, otherwise none of the members of the main or auxiliary trusses showed any indication of failure before falling.

The Columbia Road wall was about 5 in. out of plumb at the main bearing, shattered at the southwest corner where an auxiliary truss rested upon it, and cracked practically through its height in three other places. This wall pulled away from the stage wall about 2 in. at the top and pulled that wall apart, about halfway between the stage and the Columbia Road wall.

The main mass apparently fell free of the Columbia Road wall as no scratching is evident, but the four auxiliary trusses on the east side did scar the east wall with their bearing ends during the collapse. The flanges of the bearing beams which carried the wall end of these trusses were bent downward, with the exception of the northernmost which is practically straight, and the southernmost which carried a truss independent of the main truss.

The slab was badly shattered by the fall but in the southwest corner it remained suspended without rupturing over a span of about 20 ft.

Eye witnesses testify that a crack was seen to open in the plaster up near the left of the stage and that an instant later the roof fell. A witness who sat in the balcony near the column of the main truss testified that there was no noise or disturbance in that vicinity before the crack at the stage was noticed or before the collapse occurred. The testimony given shows rather conclusively that the initial movement, regardless of its cause, occurred in the vicinity of the west side of the stage.

Investigations were made by several committees: the Washington, D.C., chapter of the American Institute of Architects; the Associated General Contractors; the American Association of Engineers; and joint Army and Navy Engineers. All told, the list of items of faulty construction was a formidable one and probably many, if not all of them, were contributing causes of the failure. Probably no single fault would have been sufficient to have caused the collapse.

The first of the reports to be published was that of the joint Army and Navy Committee, and from it the Associated Press abstracted the following:

Twenty one evidences of weakness in the structure.

1. Use of tile wall to carry heavy concentrated loads.
2. The unusual height and length of the unsupported tile wall at Columbia Road.
3. Lack of a column with knee brace or pilasters under the main truss of the wall.
4. Inadequate bearing of beams on the tile wall with small bearing plates, regarded as particularly important in connection with the skew bearings.
5. Absence of concrete fill in the tile at beam bearings with wall bearing plates on edge of tiles.
6. Insufficient anchorage of steel to walls.
7. Roof slab too thin for span and loading, and insufficient steel reinforcing.

8. Absence of restraint at sides of roof slab.

9. Ceiling load concentrated in center of roof slab.

10. Short bearing of main truss on lintel, and truss not stiffened for bearing.

11. Absence of bracing between steel members, and particularly at ends of main truss.

12. No stiffness in connection of main truss to column.

13. Use of bolts instead of rivets in all field connections and insufficient bolts at truss connections.

14. Column top splices not milled.

15. Reinforcing steel in slab was not raised over the beams to take negative bending over support.

16. No evidence of increase of reinforcement in end panels of slabs.

17. Heavy load of 9½ in. of cinder concrete fill at center of drainage ridge.

18. Loads of ventilating equipment where cinder drain occurred and absence of special bracing to take vibration.

19. Defective details as witnessed by evidences of splicing for chord and end post cover plates.

20. All tile not hard burned and some have scant webs and walls, notably at top of Columbia Road wall.

21. Insufficient bonding between front and back tile in Columbia Road wall.

The report of the investigating committee appointed by the Associated General Contractors of America was probably given more publicity in architectural and engineering magazines than that of the joint Army and Navy committee. The following summation of causes is taken from this report:

The collapse of the Knickerbocker Theater roof is attributed by this committee to the unseating of the north end of the main truss, primarily through failure of the northernmost top gusset plate of that truss and the bottom chord where it rested on the Columbia Road wall. This ½-in gusset plate, which furnished the only lateral stiffness of a heavy upper chord, buckled to the west, and the lower chord, where it rested on the wall without reinforcement for bearing stresses eccentrically, was crushed and tilted under a combination of tension, compression, and torsion—or twisting stress. This action, which is later explained in more detail, culminated from many weaknesses of design, the most vital of which were as follows:

First: The overstressing of steel roof members which produced excessive downward deflection of the main truss. This overstressing resulted from (a) using loads for the design which were greatly exceeded by the actual load, (b) from reducing the height of the main truss, as originally designed, and (c) from failing to splice the top and bottom moment plates which were placed on the truss in short sections and did not butt together.

Second: The designing of the main truss bearing so that the center of stresses fell on the edge of the wall, thus producing eccentricity and tension in the bearing.

Third: The failure to anchor trusses to the walls especially over a short skew bearing at the wall support of the main truss. The design failed to provide for the additional outward thrust against the wall which developed at such a bearing when the truss was deflected.

Fourth: The failure to reinforce the bottom chord of the main truss where it bore on the wall or even to connect the channels of this chord where they extended beyond the gusset plate to form the bearing. Thus no adequate provision was made in the bottom chord for three additional stresses acting upon it, over and above the load which it carried as a member of the truss. These stresses were (a) the full load of the north end of the truss, (b) a twisting caused by the skew bearing, and (c) the tension caused by the center of stresses falling on the edge of the bearing instead of at the center.

Fifth: The omission of a cover plate at the junction of the top chord and batter post of the main truss, thus limiting the stiffness of the truss to that of ½-in. gusset plate at that point.

Sixth: The placing of a heavy concentrated load on a high and relatively thin hollow tile wall, neither adequately braced nor integrally bonded and further weakened by many openings and chases.

Seventh: Lack of diagonal or other bracing throughout the steel structure, and especially around the two columns carrying roof trusses, thus providing insufficient rigidity to the structure as a whole.

One or possibly all of these weaknesses might have been present in the building without causing collapse had their effects been distributed to various parts of the structure, but unfortunately it occurred that these effects combined and concentrated at the north end of the main truss.

Several other theories for the failure were advanced, some of them deduced from evidence taken at the coroner's inquest. Eyewitnesses testified that the first evidence of the failure was a plaster crack that appeared in the proscenium wall near the Columbia Road wall. The top of this 40-ft-high, relatively thin wall had moved outward 5-in. at the point of bearing for the main truss, and there was no connection between the steel truss and the single beam that formed the grillage for the truss. There was some evidence that this wall had been pushed outward by expansion of the truss, that mortar debris had fallen in so that when contraction occured, the truss slid on the grillage without pulling the wall back, then moved the top of the wall again a small amount when a new expansion took place. This movement, repeated nobody knows how often, combined with deflection of the truss, a weakened condition caused by the various circumstances outlined in the several reports, and the skewed bearing at this point were undoubtedly the last straw.

A comparatively large amount of space has been devoted to this failure because the engineer, the architect, and the contractor may learn so much from the collapse and from the various reports that have been made available.

Among the suggestions made in the various reports following the failure were: (1) the charging of sufficient fees for building permits to make possible an adequate force for checking and inspection of new construction; (2) provision by the owner of proper and adequate plans, specifications, and inspection on the part of the architect and engineer; (3) revision and standardization of building codes; and (4) registration of architects, engineers, and contractors by the city building department.

Darlington Hotel—New York, N.Y.

One of the most flagrant violations of all rules of safety and common sense in cutting costs in building construction came to light in the collapse of the Darlington Apartment Hotel [96] in New York, N.Y., March 2, 1904, killing 21 persons.

It was a skeleton-type apartment hotel 55 ft wide and approximately 100 ft deep from the street, and was to have been a twelve-story-and-penthouse building. From the first, everything about it was speculative—a plan to make a quick dollar and get out from under. A scathing article in the *Engineering News* stated that

> A number of men with no knowledge of building or engineering organized a company to engage in real estate and building operations. They engaged a firm of architects to prepare plans and to secure building permits for a structure to cost about $350,000. The company worked purely on speculation, intending to sell the building at a profit as soon as possible after its completion.

Up to this point of course, no wrongdoing is indicated.

The company employed the architects for a total fee of $1,200—about 0.3 per cent of the proposed cost of the job. Actually, they only bought a set of plans—no supervision. The AIA fee for this service today would be approximately $16,000.

Neither the architects nor any of their assistants nor employees were engineers. Bids were taken on the plans prepared—the bids ran high—and one of the bidders prepared a new set of structural plans, using lighter material to cut the cost. At the request of the owner, one of the architects signed the plans so they could be filed for permit, although the architect testified that he had not checked the plans. These plans, as submitted to the building department, were disapproved. The contractor then prepared a third set of plans and proceeded with the construction under this set of plans without even asking approval of the building department. At that time, the New York City Building Code had no provision for stopping such work.

The plans as prepared, and the structure as built, had brick curtain walls supported on a skeleton framework of steel beams and cast-iron columns with concrete arches between the beams—a common form of construction at that time. The beams were designed to carry a live load of 60 lb and a dead load of 60 lb. The steel details were fairly regular with connections bolted throughout—both shop and field connections. The square cast-iron columns ranged from $6 \times 6 \times 3/4$ in. up to $10 \times 10 \times 1\frac{1}{4}$ in. and were in accordance with standard practice of the time, except that many of the columns were detailed so that they carried a heavy eccentricity, the load being carried off the column center. Foundations were either granite block or concrete—the news articles are not clear which. Common practice in those days was to use granite block capstone and spread the load to the soil through plain concrete with no reinforcement, and it is probable that this was done.

The building was located in the middle of the block with four-story residential buildings on either side of it, and beyond it on the next street, separated only by a narrow courtyard, was the eleven-story Patterson Hotel.

From the beginning of the construction work, in the summer of 1903, complaints were filed by the building inspector—complaints about the foundations, the construction methods, loose connections, and the like, but little or no attention was paid to them by the owner or the contractor. The secretary of the owning company, a lawyer, stated that he had not hired the architect to supervise the work because the question of safety was up to the building inspector. In August 1903, the building inspector reported that some of the foundations were on rock and others on soil—the rock sloped off toward the rear. He recommended—apparently he had no authority to order it—that larger footings be used where they rested on clay. The owner of the Patterson Hotel, directly behind the Darlington, stated that quicksand had been found under the east and south walls of his building and that, as a result, his foundation work had cost an extra $10,000.

When two stories of the framework were in place, the owner learned that the steel contractor was working from plans of his own, but he took no action except to tell the contractors that "they had a frightful nerve to draw extra plans without our consent or that of the building department." He made no attempt to notify the building department, however. Apparently, about this time, he began to lose confidence in the contractor and he employed a draftsman from the architect's office to check every piece of steel as it arrived. This man had no experience in steel construction and had never gone higher than the third floor of this building. There was no single person responsible for the construction work, and the steel contractor did as he pleased.

A week or two before the failure, when the ninth floor steel was in place, the structure was found to be 18 in. out of plumb. It was racked back into place with block and tackle, and temporary bracing was installed at that time. Throughout the entire period of construction, the framework swayed noticeably. The bolt holes in the cast-iron columns were cored instead of drilled as required by the code. The holes were cast into the column, with all the irregularities inherent in this type of work. Materials were not stored carefully, and no thought was given to the possibility of an overload. On the morning of the failure, the building inspector filed a report of "A violation of section 41 of the building code as follows: in that of constructing the side walls more than two stories in advance of front wall, and the floor beams were not properly bolted and tied."

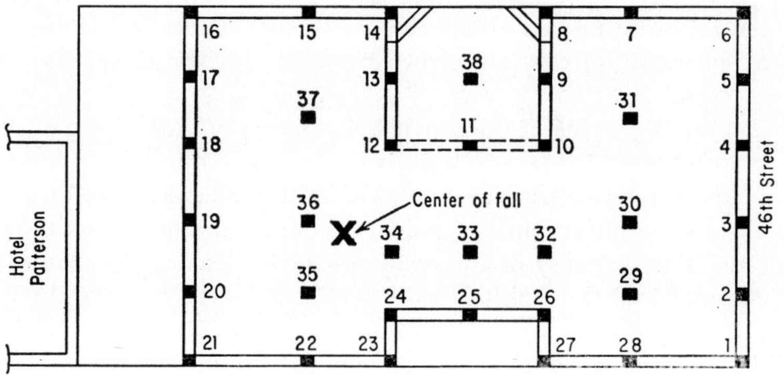

FIG. 7.6. Basement column plan of Darlington Hotel

On the day of the collapse, the tenth floor steel had been erected and the concrete followed within two or three floors. The front wall was just begun. The derrick for hoisting material was on the front bay, and the steel for the structure above the tenth floor was stored on the tenth floor framing. Just before the collapse, observers from across the street saw the frame sway back and forth, then apparently fall inward. With a few exceptions, everybody in the building was killed. An examination of the wrecked frame showed that the steel beams and girders, although badly twisted and bent, were not pulled apart too badly. A few connection angles were torn but, for the most part, the bolts were broken or stripped. The cast-iron columns were almost all broken in some way. The smaller columns were broken across the body, and the larger ones broke in the flanges connecting them. The lugs to which the beams were connected were broken off.

The technical reasons for the collapse are summarized in a report published in the *Proceedings of the American Society of Civil Engineers,* April 1904, by H. deB. Parsons, who had investigated the failure. (Details are shown in Fig. 7.6–7.10.)

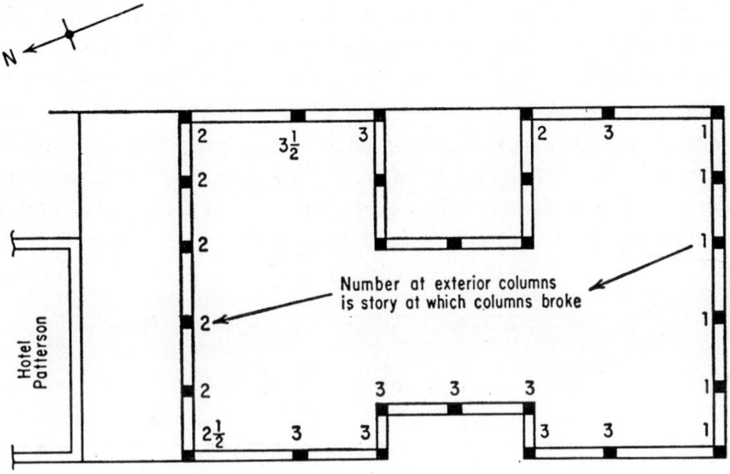

FIG. 7.7. Effect of collapse on wall columns of Darlington Hotel

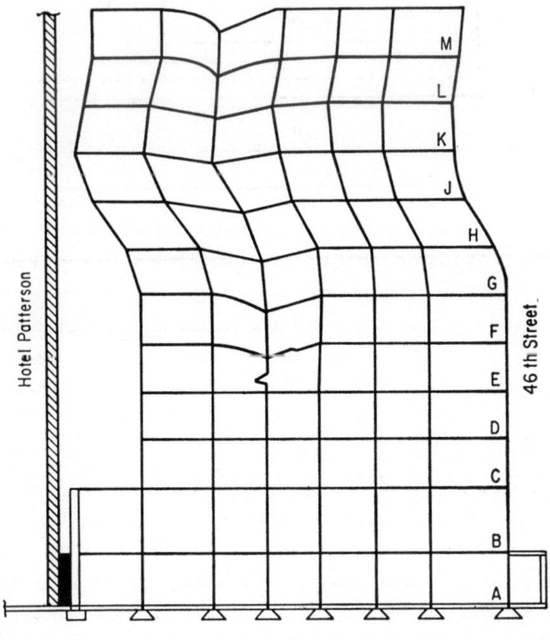

FIG. 7.8. Collapse of Darlington Hotel—vertical section north-south during failure

The bolts holding the beams and girders to the column lugs were of smaller diameter than the holes, so that the columns received little or no lateral support. All of the loads were eccentrically supported on side brackets. The columns were too long to carry the superimposed loads. One column, located at or near the center of failure, broke and precipitated the collapse—all falling

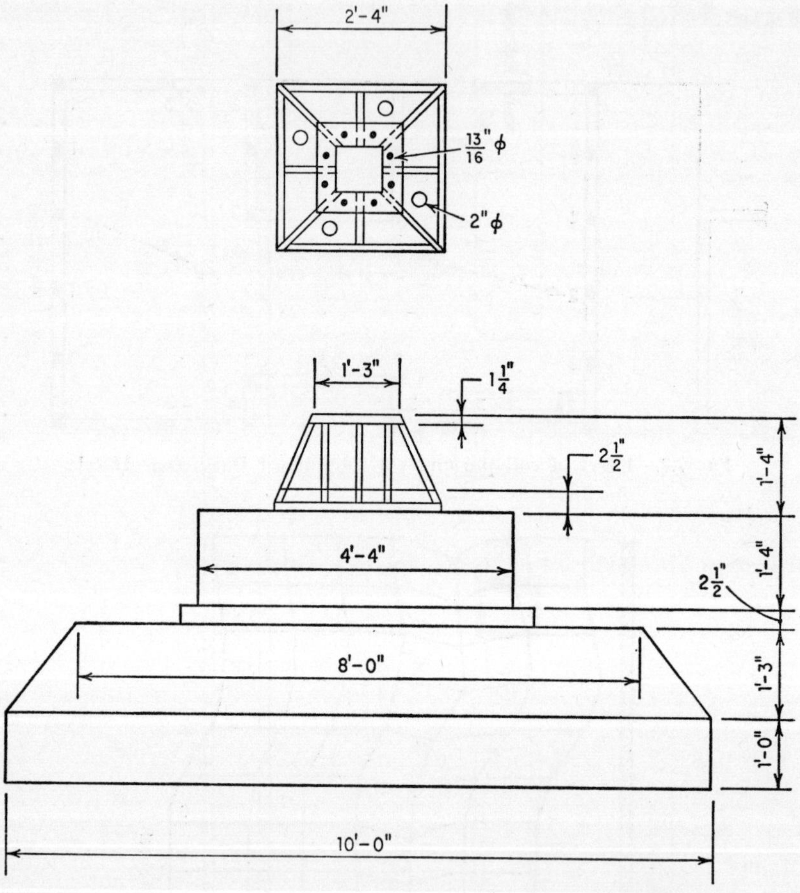

FIG. 7.9. Typical column footing of Darlington Hotel

in toward the center. The structure collapsed because of lack of lateral support for the story columns, permitting them to act as long columns with a slenderness ratio (l/r) in excess of known limits of safety.

A coroner's jury, composed almost entirely of architects and contractors experienced in building work, immediately began a thorough investigation and three weeks later reported: "We find that the collapse was due to defective designing, a total absence of competent supervision, and the employment of inexperienced contractors. We further find that Eugene F. Allison (the owner) and the firm of Pole and Schwandtner (the steel contractor) were grossly, criminally negli-

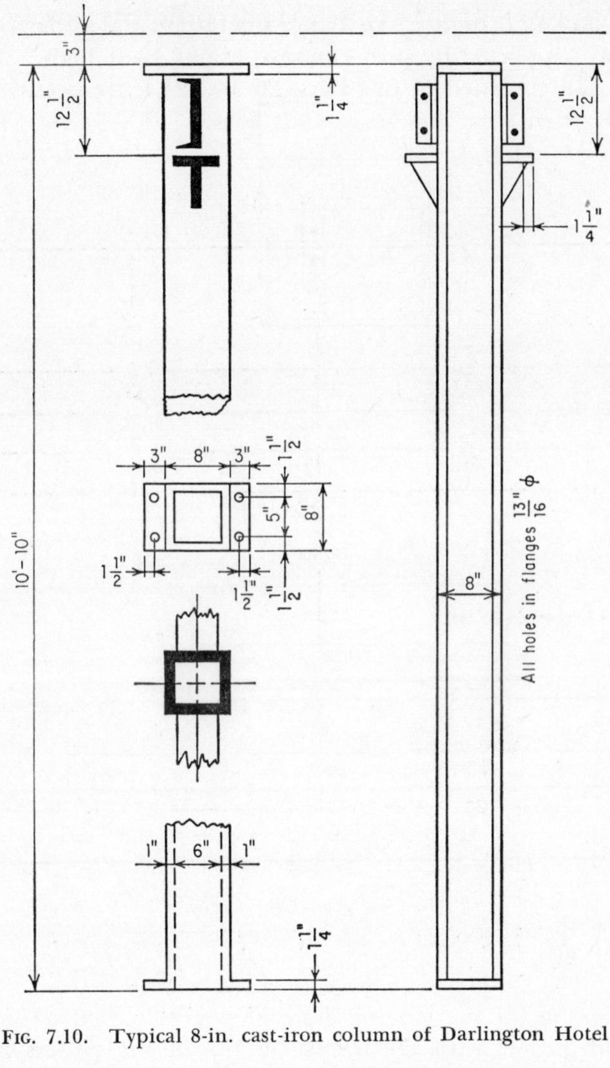

FIG. 7.10. Typical 8-in. cast-iron column of Darlington Hotel

gent in the methods adopted and followed by them in the erection of said building."

Bronx coal pocket—Bronx, N.Y.

Professional responsibility took on real meaning and emphasis when the designer of a faulty coal pocket was convicted of manslaughter.

The 3,000-ton coal pocket [97] on the bank of the Harlem River, Bronx, N.Y., collapsed on February 21, 1931. The collapse resulted in the death of three men and the injury of eleven others.

The coal pocket consisted of a timber bin 30 × 125 ft in area and between 35 and 40 ft high, carried on a steel-beam-and-girder framework, which was supported by steel columns about 10 ft high. The weakened portion consisted of two bays, 20 and 24 ft, respectively, a total of 30 × 44 ft in area, and three 10-ft bays.

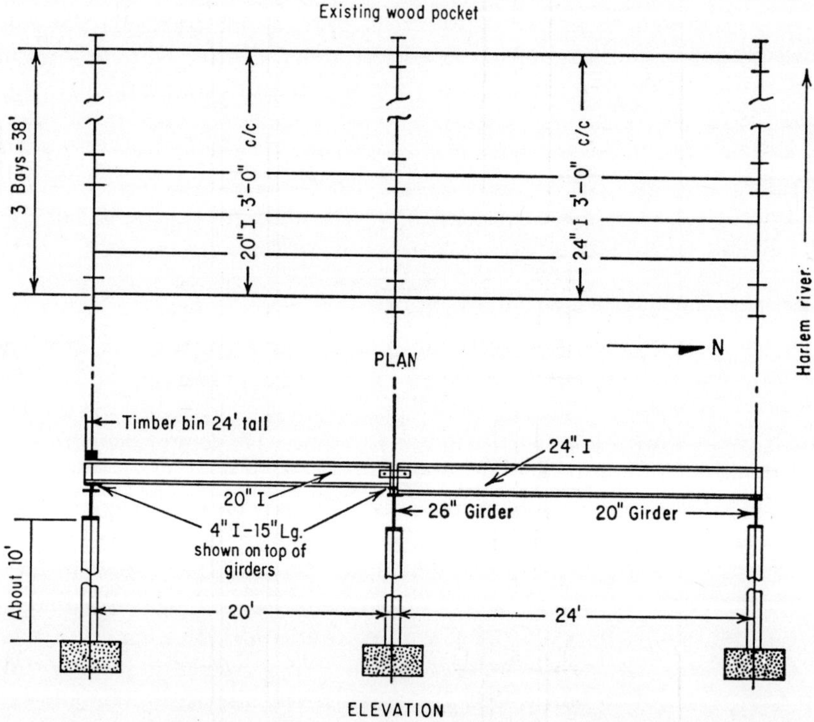

FIG. 7.11. Steel framing in two bays of coal pocket

The framing for the base of the coal pocket was made up of steel beams and girders with the beams on top of the girders instead of being framed in. The floor beams for the 20-ft bin were 20-in. beams. To bring the beams to a level floor elevation, 4-in. H-beams, 15 in. long, were placed on top of the girders under the 20-in. beams. This layout was used to enable the steel contractor to use steel he had in stock. (Fig. 7.11). There was no bracing in the vertical plane, and the failure was blamed on this fact, along with the buckling of the web of the 4-in. filler beams. At the time of the failure, one bin was full and the other bin was being filled.

The conviction of the designer was appealed and the following points were brought out.

1. That the structure was built without any building permit. The Bureau of Docks, whose duty it was to issue permits for the waterfront properties, had not issued the permit because of legal questions involving property rights. On the other hand, this authority had not stopped the work on the structure.

2. It was built without any inspection or supervision.

3. The designer's advice to build on pile foundations was not followed, and, in the original trial, all testimony relative to foundations was ruled out despite the fact that one foundation had settled 4½ in. and tilted.

4. The steel erection had been done in a careless and slipshod manner.

The Court of Appeals granted a new trial, but, after some delay, the Bronx District Attorney decided to drop the case entirely.

Apartment building—Rio de Janeiro, Brazil

It has been said that the building code in Brazil is good, but evasion is easy and the building inspection is completely inadequate. The author is reminded of an occasion, some years ago, when he was called upon to do work in a small neighboring city. He telephoned the superintendent of public works—there was no building department—to inquire if the city had a building code. He received the reply: "Yes, we've got one, but nobody around here seems to know where it is." This seems to apply around Rio de Janeiro, Brazil, where, in February 1946, an almost completed ten-story apartment house [98] in Copacabana, Rio's fashionable beach district, collapsed and killed ten men.

Examination of the site before construction showed the presence of an underground stream about 16 ft below the surface. The stream was pumped while foundations were being poured, but no piles were driven. The foreman in charge of reinforcing steel was discharged after four or five months because he called the attention of the contractor to defects in the steel. There was almost no control exercised over the concrete mixing, and the mix was not of good quality.

On the day of the failure, the head of the construction firm was shown a crack in a second floor column. He ordered that a shore be placed to relieve the load and went to get some material to fill the crack. What material could have saved the job at that point is problematical. At any rate, when he got back, the material was unnecessary. The building had collapsed.

West Jordan School—Salt Lake City, Utah

It is appalling to realize how many people are living and working every day in a potential death trap. The West Jordan School,[99] a reinforced-concrete building, near Salt Lake City, Utah, had been in constant use for 22 years when alert teachers noticed plaster cracks in the ceiling and emptied the school before a section of the roof fell, in 1947. As a result there were no casualties.

The inspection following the collapse disclosed alarming conditions. The workmanship and materials were both poor. Pieces of wood, brick, and old mortar were found in the concrete at critical points of the concrete construction—any of which could have caused failure long before it occurred.

The brick bearings of girders were not designed or built heavy enough to carry their loads. Reinforced-concrete girders were underdesigned and were built even smaller than shown on the plans. Gravel had stones in it too large for the members where the concrete was used.

As an aftermath of the failure, the state superintendent of education requested all district boards of education to employ experienced engineers to inspect all new buildings, particularly where the use of reinforced concrete was involved.

Majestic Dance Hall—Salt Lake City, Utah

Given a design of questionable stability, adverse conditions throughout construction, insufficient or no maintenance, and a dance hall used also for a roller skating rink, and one has all the elements of a disaster. Only the fact that no one was in the building at the time prevented this accident from being a catastrophe, when a heavy snowfall, most of it in a drift on one side of the roof, collapsed a number of 110-ft wooden Howe trusses over the 150 × 250-ft Majestic Dance Hall,[100] in Salt Lake City, Utah, March 25, 1916. The frame building was about two years old.

The construction of the building was a rush job and was actually ahead of the plans. The lumber was green and not properly seasoned. The diagonals were wood members, mortised into the chords, and the verticals were steel rods (Fig. 7.12). Apparently, there was no lateral bracing. The lumber had apparently shrunk before the building was finished and shims had been driven in to tighten the mortised joints. The chords were built up of 2-in. dressed members, spliced together, and there were places where two of the seven 2 × 14's in the bottom chord broke joints at the same point, leaving an effective joint of about 70 per cent of the member.

When completed, the trusses had been ceiled in completely, making an ideal condition for shrinkage and warping of members. Electric lights had been installed in this ceiling space for inspection purposes, but the bulks were all missing. No attempt was made to tighten up the steel rods, and the top chord sections had taken a serious sag.

During the time this deterioration was taking place, the owning company went through bankruptcy and no maintenance work was done. The inevitable failure took place, apparently by failure of the top chord through buckling. From experience with similar failures, the author would suggest the possibility that, with the play in the bolt

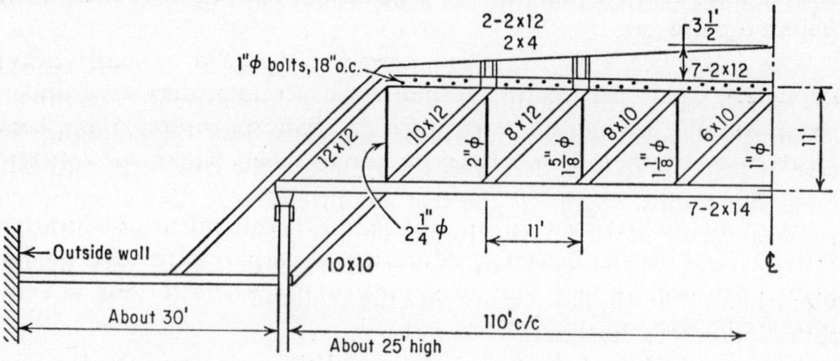

FIG. 7.12. Elevation of 110-ft-span Howe roof truss

connections due to the shrinkage of the wood, the top chord member was probably acting as seven individual members, rather than as one single solid member.

Construction defects in a concrete building

The name and location of the building written up in this section was not given in the article from which it was taken—an article by Grayson Gill of Dallas, Texas, in the *Engineering News-Record* of May 10, 1928. It is the story of a well-designed building where the contractor apparently went out of his way to do a poor construction job.

A four-story, reinforced-concrete joist-and-beam garage [101] had been built in 1925—designed for the future addition of eleven extra stories. In 1928 it was decided to add the extra stories but, since the investigation of the bedrock and caissons indicated their ability to take a greater load than had originally been planned, it was decided to strengthen the columns and to add fourteen stories instead of eleven. None of the people who undertook the work in the addition, architect, engineer, or contractor, had had any connection with the earlier job.

Preparatory to undertaking the work, a visual inspection was made of the building, and alarming conditions were disclosed. Construction joints in concrete beams had been made near the supports where the shear is greatest—the worst possible point. Some of the construction joints closely approximated the line of failure by diagonal shear, and when a passenger car was driven over the garage floor, the joint opened perceptibly. In two 12 × 24-in. beams the vertical wooden bulkheads for the forms had been left in the beam at a point within 3½ ft of the support, thus establishing a plane of cleavage. In many places reinforcing bars were so crowded together in the bottom of the beam that no coarse aggregate could flow between them, thus giving a weak bond and allowing only the cement and sand to form the underside of the beams.

Where the wall beams framed into the columns, the closely spaced beam reinforcement combined with the column reinforcement to effectively dam off the flow of coarse aggregate into the column. In several places 3- to 6-in. deep voids were left in the core of the spirally hooped columns. Some of the bad conditions were discovered when the contractor started removing obviously defective work with an air hammer and loosened large areas of other defective work by the vibration. The hollow spots in the columns were detected by sounding with a hammer.

An investigation of the foundation conditions to verify the location and condition of rock, however, disclosed really hazardous conditions. The depth and condition of rock were found to be in accordance with reported facts. The foundation caissons, however, were far short of what the original drawings required. The square concrete caisson had not been brought up to the underside of the floor to receive the column but had been terminated 3½ ft below the floor. A batter-faced plinth had been built on top of the caisson to the floor line. The first plinth uncovered did not center on the caisson, but actually overhung the top of it as shown in Fig. 7.13, so that the center of the column was 20½ in. off center from the caisson. There were no dowels between the plinth and the caisson, and a layer of smooth laitance (a whitish scum) effectively destroyed any possible bond between the two. A test pit at this point disclosed that the caisson, instead of going to bedrock as had been certified by the builder, stopped on a soft white chalk formation 7 ft above bedrock. Similar conditions were found throughout the building so that before placing the additional load on the columns fifteen caissons were removed and replaced with good caissons to rock.

The concrete in the caissons indicated that it had been dropped into the water, thus separating the aggregates from the cement and permitting the infiltration of clay into the concrete. Laitance as much as 10 in. thick, containing clay balls and other debris, was found at the top of the caissons; below the laitance, a section of soft,

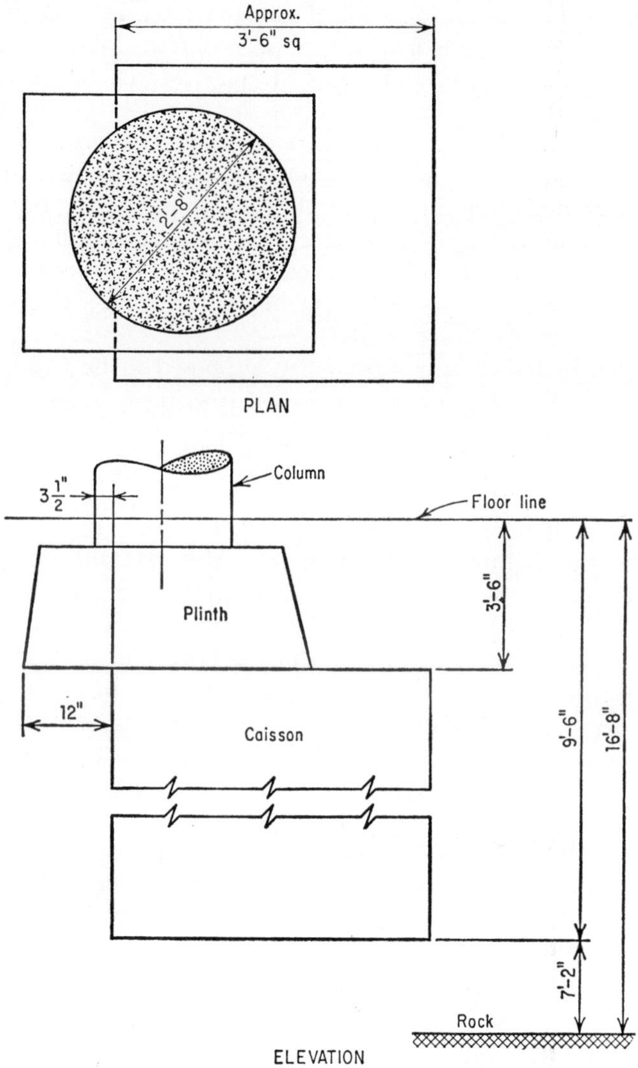

FIG. 7.13. Detail of footing for concrete building

sandy concrete occurred, and in the bottom was 3 to 3½ ft of coarse sand and gravel containing practically no cement binder, and which could be shoveled out.

In spite of all the faults of the original construction, the work of making a sound eighteen-story building out of a sick four-story structure was successfully accomplished.

Partially constructed buildings—New York, N.Y.

By a strange coincidence, freak of nature's laws, or what have you, six buildings[102] under construction in different locations in New York, N.Y., collapsed on Sunday and Monday, March 19 and 20, 1905. Others followed later. The failures of the first five—all on Sunday—occurred at 9 A.M., 2 P.M., 6 P.M., 11 P.M., and 11:30 P.M. The sixth, which was adjacent to and continuous with the first, fell at 6 P.M., Monday. Nobody was in any of the buildings at the time of the failures. All of the buildings were five- or six-story, flat buildings of brick wall and wood floor construction, and all had been completed either to the roof or to the tier next below the roof. The buildings were as much as four miles apart.

The cause of the failures was at first supposed to be frozen mortar in the brick walls. Following a period of freezing weather, Saturday—the day preceding the beginning of the series of collapses—had been warm, with the temperature reaching 70°F, and Sunday was warm and rainy.

The board of experts that investigated, however, reported that in every case the failure had been directly due to faulty foundation walls: full of voids, built of irregular and undersized stone, and improperly bedded or bonded. The sudden change in temperature may have been a contributing cause to the failure of the foundations, but the walls of all of them were of such inferior workmanship and material as to make them unsafe. The building code had been extensively violated, plans on file had been disregarded, and changes ordered by the building department had never been made.

J. L. Burns and Co.—Northfield, Vt.

Strange as it may seem, what may appear to be one of the simplest types of framing, and what was, for many years, one of the most used for roofs—a wooden roof truss—may involve more application of theory and more constant inspection and maintenance than some other apparently more difficult types.

This fact is proven by the failure of a large granite shed[103] belonging to J. L. Burns and Co. at Northfield, Vt., on February 26, 1910. The collapse resulted in injury to two men. The building was designed and constructed by an old bridge foreman with no knowledge of stresses in structures. It is only fair to say, however, that it is possible that at the time of the failure the roof was carrying a 50-psf snow load.

The shed was 256 ft long and 50 ft wide, and the roof was carried by a series of simple trusses of the type shown in Fig. 7.14. The failure probably took place in the bottom chord splice, either by the shearing on the horizontal plane at the bottom of one of the keys,

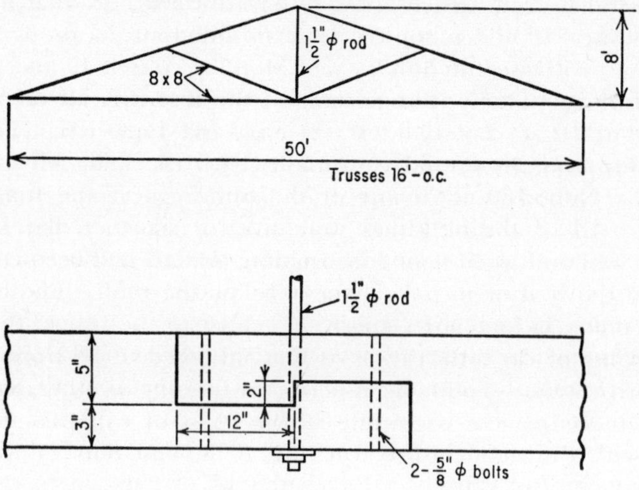

FIG. 7.14. Tabled fish plate joint for granite shed truss splice

or by crushing of the 2-in.-deep bearing of the keys. A computation shows that the truss is weak in both of these factors for the load it was indicated as carrying.

Concrete silo near Buffalo, N.Y.

A normally built silo is a tall, cylindrical storage structure of steel, concrete, masonry, or wood, built either as a single cylinder or as one of a group of such structures, attached to each other. Whatever the material to be stored may be, either liquid or solids, it exerts a lateral pressure against the sides of the cylinder, and the structural safety is dependent upon the ability of the walls of the cylinder to withstand this lateral pressure. This lateral pressure is tied in by means of hoops—similar to barrel hoops, which prevent the barrel staves from buckling outward. In a steel silo, the entire wall acts as a hoop. A concrete, masonry, or wood stave silo, however, is tied by steel hoops. In reinforced-concrete silos, the hoops are buried in the concrete, which serves merely to contain the material and to carry the load into the steel hoops. It is obvious, therefore, that the steel must be in the form of a continuous hoop, or, if in sections of the arc of a circle, they must be lapped into the concrete far enough to develop the full strength of the steel in tension through the bond of the concrete to the steel.

In June 1954, a reinforced-concrete silo about 21 ft in diameter and 45 ft high was used to contain a clay slurry—a mixture of clay and water weighing approximately 80 lb per cu ft. The slurry had full fluid pressure laterally—that is, a lateral pressure on the bottom foot equal to the total weight of a column of the slurry 1 ft square and the full height of the column. An agitator mounted on the top of the silo was used to stir the slurry.

The steel hoops were strong enough to contain the lateral pressure of the material, except that in reusing a previous design, someone decided to install an access door near the bottom of the silo. This is frequently done, but the hoop stress through the zone of the door is contained by building a door frame of sufficient strength to carry the stress around the door and by welding the interrupted bars to the frame. In this instance, the bars were placed with no connection to the frame. Even if they had been welded to the frame, the frame was not designed to carry the stress that would have developed.

About two months after the silo was put in operation, the door and frame and the attached concrete blew out with explosive violence, rupturing the concrete in this zone and collapsing the tank, injuring fourteen men, and causing a large amount of damage to the plant by flooding the power plant with liquid clay.

The section which blew out was about 4 ft high and 3 ft wide, and the total pressure it bore was about 37,000 lb. This pressure, carried by the concrete alone without the aid of the steel hoops, would bring about a tensile stress of more than 300 psi in the concrete—a stress in excess of what even good concrete can carry.

Several other questionable factors were suggested as contributing causes, such as an insufficient lap of the sections of hoops, and the surge action in the tank caused by the agitator on top of the tank. The manner in which the failure occurred, however, indicates that the collapse was caused by failure to provide for tieing the tensile stress around the door that was installed near the bottom of the silo.

7.2 Tank Tower Failures

Tank tower failures are so similar in their cause that it is perhaps repetitious to write up so many of them, yet their recurrence is in itself reason for giving more than one example. Those which are written up here are only a few of many that have happened through the years and continue to happen.

Two of these failures were definitely caused by the removal of essential bracing; the third by failure to stiffen the top flange of the supporting beam properly, and the fourth by either failure of the supporting beam or weakness of the supporting wall, or both.

Fuller Brush Co. Tank Tower—Hartford, Conn.

At the time of the failure of the Fuller Brush Co. tank tower,[104] at Hartford, Conn., March 31, 1923, the author was working on a similar tower for an industrial plant then on the drawing boards. Without knowing what had caused the collapse, the insurance company immediately demanded drastic revision of the unfinished plans, thereby indelibly impressing this failure on the author's mind.

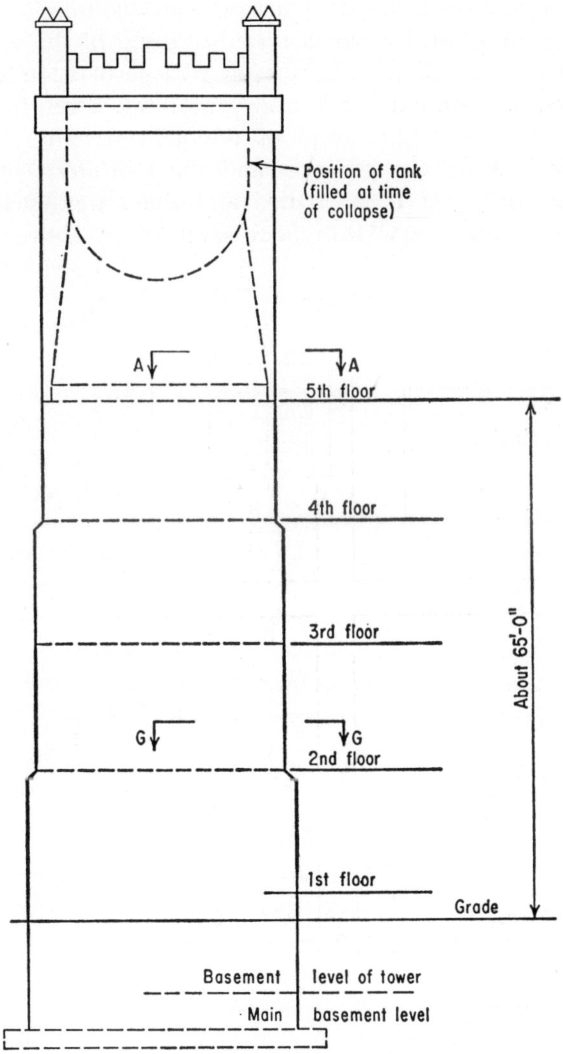

FIG. 7.15. Elevation—Fuller Brush Co. tank tower

The tank and tower that failed were part of a new building being erected for the Fuller Brush Co. The main portion of the building was three stories high and had as a central feature, a 28-ft-square tower which projected several feet past the front of the building (Fig. 7.15). The tower was five floors high, the fifth floor built to contain a 50,000-gal steel water tower. The tower was supported on four inclined plate-and-angle steel columns, which were braced about 1½ ft above the fifth floor level by means of 1-in. square rod diagonals with turnbuckles and tied together at the base by means of a square of 4-in. channels (Fig. 7.16). Each diagonal column was supported by a grillage of 12 in. I-beams across the corner of the square, which, in turn, was carried by a wall-bearing grillage of 12-in I-beams (Fig. 7.16 and 7.17). Around the masonry wall was a reinforced-concrete band at the approximate level of the grillage, but there was no positive tie between the grillages and the reinforced-concrete band.

After the tank had been erected and the tower enclosed, it was decided to install a new fifth floor several feet above the original

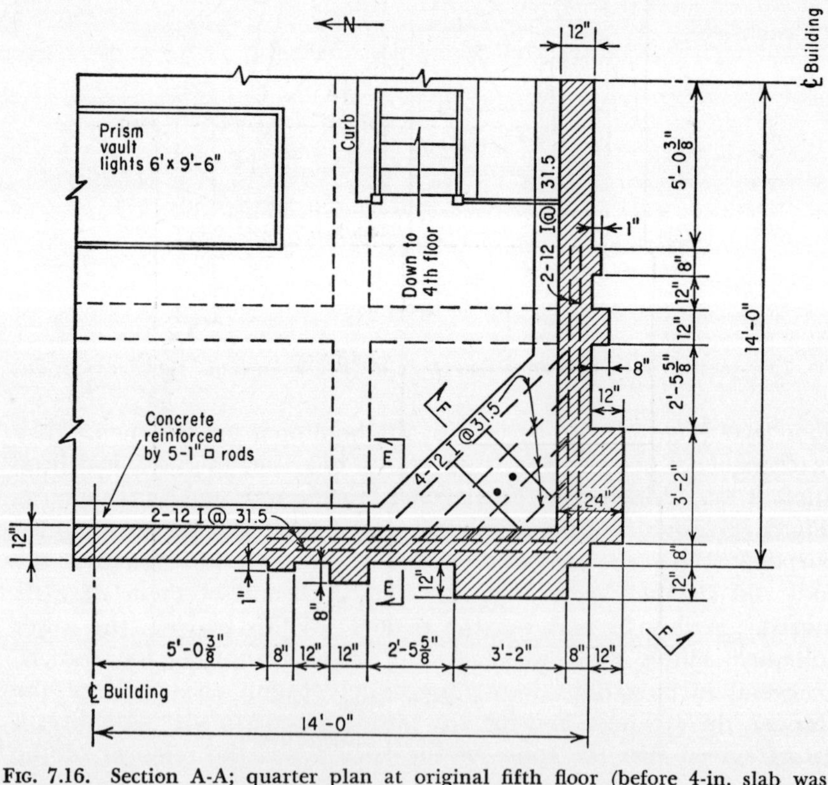

Fig. 7.16. Section A-A; quarter plan at original fifth floor (before 4-in. slab was superimposed) showing tank column supports at southwest corner of tank tower

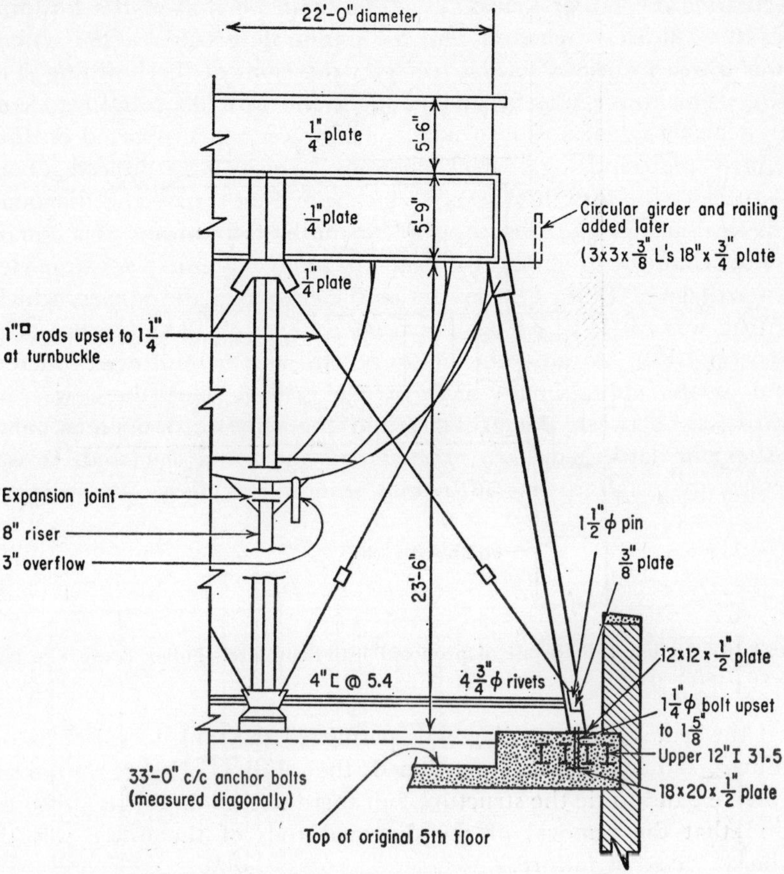

FIG. 7.17. Section F-F: half elevation of Fuller Brush Co. tank showing supports

floor level (Fig. 7.18) to be used as a schoolroom for salesmen. The exact sequence of operation is uncertain, but after the tank had been filled, it was decided that the diagonal bracing rods and 4-in. channels (Fig. 7.18) would interfere with the use of the room for its intended purpose. With the approval of the architects and engineers, the rods and channels were disconnected and several of them were removed. Within 24 hours of the removal of this bracing, the tower collapsed, killing nine workmen and seriously injuring four others.

Several investigations were immediately begun on behalf of the coroner, the architect, and the city building department. All investigators agreed that the cause of the failure was the removal of the bracing. One of the investigating engineers put forth the theory

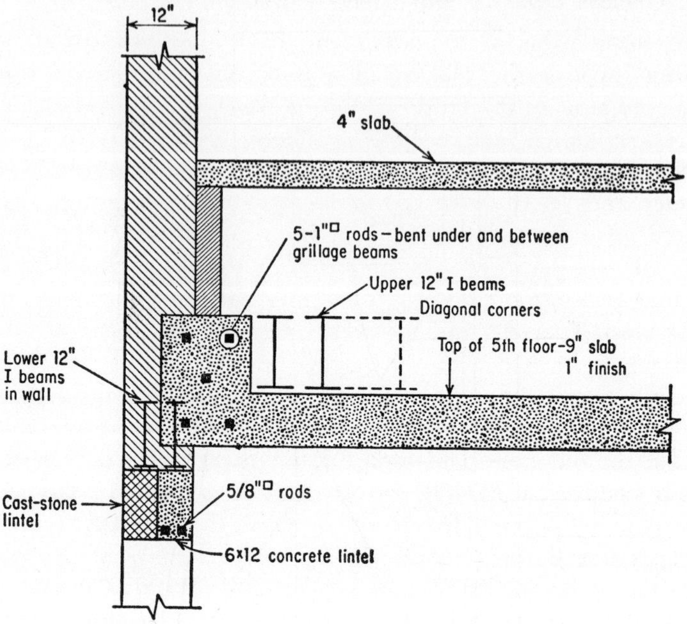

FIG. 7.18. Section E-E: detail of proposed fifth-floor level—Fuller Brush Co. tank tower

that the removal of the diagonals permitted the tank to spiral before failure. Other theories questioned the ability of the reinforced-concrete band to tie the structure, but there was no doubt in anybody's mind that the removal of the diagonal rods of the tower was the primary cause of failure.

Peabody tank tower—Peabody, Mass.

In April 1926, a 15,000-gal elevated tank [105] on a 65-ft high steel tower collapsed in Peabody, Mass., killing one man and injuring a number of others. The tank and the tower were about 20 years old and in good condition. The wind at the time of the failure was only about 10 mph.

Workmen had been cleaning and painting the structure and, in order to make parts of the structure more accessible, had removed three or four clevises that attached the diagonal bracing rods to the columns. The tank rotated a full 180 degrees about its vertical axis and collapsed. In addition to the casualties and the loss of the tank, the failure did extensive damage to surrounding buildings and their contents.

Union American Cigar Co. water tank—Pittsburgh, Pa.

The installation of a water tank over the roof of an existing building involves the placing of a potential hazard over the heads of the occupants of the bulding, and, nowadays, the insurance examiners are very much concerned about how this tank is to be carried. Apparently this was not the case in 1912. A wooden tank,[106] 12 ft in diameter and 10 ft deep, having a total filled weight of 70,000 lb replaced an older tank, which had been dismantled. The new tank was filled to a depth of 7 ft, and part of the water was being drawn off when the supports failed, dropping the tank through the roof, killing several persons and injuring others.

The new tank had been set on the old steel beams, which were set diagonally across the corner of the building on 9-in. walls (Fig. 7.19). Inspection of the wreckage disclosed that the most remarkable fact was that the old tank had remained in place. The action of a steel beam in bending, where the top flange is unsupported or not properly supported, is discussed in several other sections of this book, which report similar failures. Even if the beams had been properly supported laterally and had been of new steel, the 15-in. diagonal beam at

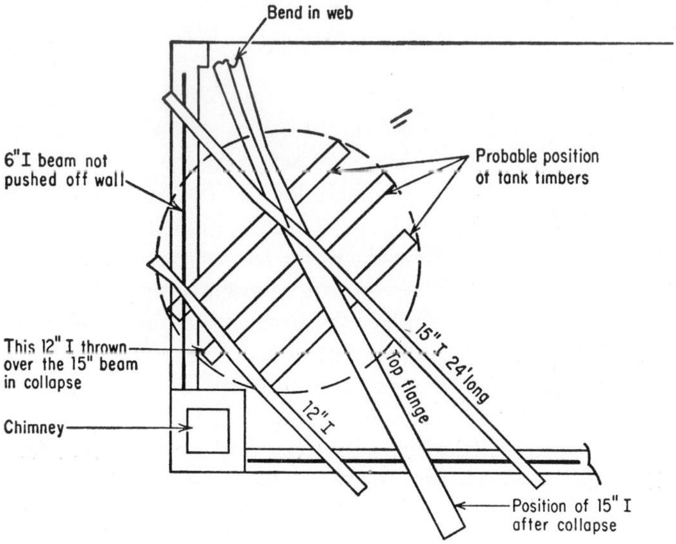

FIG. 7.19. Union American Cigar Co. water tank support

the back had only about half the strength it should have had in bending. Moreover, none of the beams, either those supporting the tank or the grillage beams, had any stiffening to prevent their tipping. The long 15-in. beam where the failure took place was badly rusted, particularly in the top flange under the tank where it most needed strength. To judge from the sag in the 15-in. beam as it lay on the roof, the stress in the beam had passed the yield point of the steel, the top flange buckled and tipped the tank sideways. The entire job lacked evidence that any engineering had even been done on the design.

R. J. Ederer Thread Co.—Philadelphia, Pa.

As stated previously, the top flange of a steel beam is of itself a compression member. In order to carry the compression, which is the direct result of bending (Sec. 1.4), the beam must be supported against lateral deflection, just as the ratio of slenderness must be kept within limits if a column is to be safe. This principle, along with the fact that even if it had been supported laterally the beam would still have been too weak, is ample cause for the failure of a water tank [107] on top of the R. J. Ederer Thread Co. in Philadelphia, Pa., December 27, 1911, resulting in the death of two men and injury to eight others, as well as wrecking a portion of the factory.

The tank, 16 ft in diameter and 16 ft deep with a total filled weight of 200,000 lb, was carried on top of a recently built fire tower, the top of which was 24 ft above the roof of the building, or 68 ft above the basement floor. Plans for the construction had never been filed with the building department, and no building permit had been issued. The tank was carried by a battery of seven 10-in. steel I-beams, which were of ample strength to carry their load. These beams were carried on top of two 12-in. I-beams 15½ ft long, which in turn were bedded in mortar on top of the brick walls of the tower. The carrying capacity of these 17-in. walls was cut to some extent by the presence of an arched opening 9 ft wide, with the top of the semicircular arch about 9 ft below the tank supports.

Investigators blamed the collapse on the failure of the arches, which pushed out the 5-ft-wide piers, but a simple computation of the 12-in. beams will indicate that the tank should have failed long before it did.

7.3 Temporary Grandstand Failures

Temporary grandstands or removable bleachers have been a fruitful source of trouble throughout the years. Sometimes, it seems as if the simpler the type of construction, the more liable it is to be done improperly. It does not seem to register with some people that engineering principles must govern the construction of simple, temporary grandstands as much as they do enormous steel or concrete stadiums. Because of the very nature and use of the structure, building departments should be more strict in examining and approving them than they are with permanent structures where the economics of the structure requires good design.

This examination should be particularly explicit for lateral bracing, connections, mud sills, and other details of such construction where the weakness is ordinarily found. The following failures will indicate this fact.

Purdue University bleacher—Lafayette, Ind.

The use of knockdown bleachers has become quite common at colleges, universities, and high schools where athletic events attract enough patronage to warrant the purchase of a type of bleacher easily erected or knocked down. Even such properly designed bleachers, however, are subject to failure, as was shown by the collapse of a knockdown bleacher [108] during a basketball game in the Purdue University Fieldhouse, February 24, 1947, resulting in the death of three students and injury to more than 1,000.

The report of the investigating committee, appointed by the governor, was printed in the official publication of the Indiana Inspectors Association and the following is extracted from this report by way of the *Bulletin of the American Institute of Architects.*

The grandstand was of the bleacher type, constructed of wood, with steel or iron connections to facilitate erection.

There were 42 rows of seats, approximately 125 ft long. The breadth (front to back horizontal measurement) was about 81 ft. The rise at the rear was about 17 ft 3 in.

The stringers were a nominal size, 2 × 8 Douglas fir. The A frames were 4 × 4's, the spacer bars 2 × 4's and the splice sills were 2 × 6's in six pieces with splices from 16 ft 2 in. to 10 ft 2 in. apart. The splice consisted of a hook fitting a U-shaped stirrup. The hook was made of $\frac{1}{8}$-in. steel $4\frac{1}{2}$ in. wide and the stirrup $\frac{1}{4}$-in. steel and 1 in. wide.

The stand was occupied by approximately 3,600 persons, the actual live load being estimated at 53 psf of the structure's area.

There was no concerted or rhythmic movement of the audience during the game.

Tests of undamaged samples of the wood and metal parts indicated the wood to be of structural grade and the metal to be of low-carbon steel or ingot iron.

The following are excerpts from the Committee's conclusions.

1. The failure resulted from the breaking or failure of the foremost metal-hook-and-stirrup-connection on the sleepers in the south central portion. The remaining foremost sleeper metal-hook-and-stirrup connections in this front line failed successively, allowing the structure to be thrust forward and downward by the load, displacing the A-frames that supported the stringers, which then broke at several places due to bending and lateral displacement.

2. The physical tests of fourteen undamaged typical sleeper metal-hook-and-stirrup connections show that their capacity was not sufficient to withstand the force the structural analysis of the grandstand showed to be acting on the foremost line of these connections.

3. The fractures of two adjacent broken sleeper metal-hook-and-stirrup connections in the region of the initial failure show an oxidized condition, indicating that partial failure had occurred at some time previous to the final collapse.

The failure started in the lower center of the stand and, by pulling the other portion of the stand toward the point of origin of the failure, caused a gradual collapse extending upward to the top of the stand and outward in both directions to the ends. Of the more than 3,500 spectators in the stand, more than 1,000 received some form of injury.

Tournament of Roses grandstand—Pasadena, Calif.

Probably the most famous temporary grandstand failure was the failure of a stand [109] during the Tournament of Roses parade in Pasadena, Calif., January 1, 1926, resulting in the death of three persons and the injury of a number of others.

The stand was 68 ft long, and 41 ft from front to back, rising to a total height of 20 ft at the back. It had 22 rows of seats and was built to accommodate more than 900 people; not that many were on it at the time of failure, however.

The framing was very simple, but with no cross bracing from front to back and only hit and miss bracing lengthwise. The report of a board of inquiry stated that the stand was "totally inadequate and insufficient." The materials used were largely secondhand and second grade or poorer; some of it was used form lumber. The workmanship likewise was of very poor quality; anchor blocks were omitted; posts were cut so that they did not seat properly on their mud sills, and so on. The Pasadena code required that such a stand be designed to carry a load of 125 psf, but at the time of failure, it was carrying less than half of this amount.

Redwood City circus bleachers—Redwood City, Calif.

Familiarity breeds contempt, even when warning of an unsafe condition is given. The circus [110] came to Redwood City, Calif., in April

1948, after a period of rainy weather that had softened the adobe soil. The city fire inspector warned the circus management that it would be unsafe to use the stands under this condition, but the warning went unheeded. When the bleacher section of eight rows of seats, accommodating 1,000 persons, was filled up, the A-frame supports with no lateral bracing sank into the soft adobe soil, and 64 customers were injured.

But the show must go on. The next day the circus opened on schedule at nearby San Mateo, when the building inspector there approved the use of 2 × 12 blocks in order to prevent the A-frames from settling again.

7.4 Specific Flaw in Design or Construction

Most of the sections in this chapter—particularly those in the early part involve failures that resulted from more than one violation of the rules of good practice in building construction. This section describes a few failures where specific error in judgment in design, manufacture, or construction brought about collapse. None of these failures resulted in deaths or injuries.

Home Theater—Chicago, Ill.

The fact that the failure occurred during the morning when the building was unoccupied probably prevented a catastrophe, when the roof of the small, one-story Home Theater [111] in Chicago, Ill., failed, Sunday, December 9, 1912. The building had been reported as dangerous, and, on Saturday morning, the building department had issued an order to close the theater, but the order did not reach the police department in time to be served until after the collapse had occurred.

The failure was apparently in a cover-plated, 24-in. I-beam on a 48-ft clear span which had been built up from two shorter pieces with web and flange splices. Based on the dead loads on the beam at the time of failure, the girder was stressed to more than 44,000 psi, which was beyond the yield point of the steel of that day—a sufficient cause for the failure.

Wilson Avenue Reservoir roof—Pasadena, Calif.

The failure of the roof over the Wilson Avenue Reservoir [112] in Pasadena, Calif., March 25, 1917, brought to light an inherently weak construction that had apparently been used for a number of similar reservoirs.

The reservoir in question had been built in 1909–1910 and was 414 × 238 ft in area and 17 ft deep, with the sides on a 1:1 slope. The lining was an unreinforced-concrete slab. The roof was made up of rough 1-in. boards on 2 × 8 joists, 16 ft long and 4 ft on centers. The girders on which these joists rested were made up of double 2 × 10's, 15 ft 10 in. on centers. The columns were 2-in. standard pipe dipped in asphaltum, capped at the bottom and set into a short 6 × 6 bolster on top. Thus there was no resistance whatever to any lateral thrust.

The support at the wall was just as unstable, consisting of a 2 × 6 redwood sill, which actually had only 2 to 3 in. bearing on the concrete and was tipped upward on the outer edge. The end of the joist was carried on a 4 × 4 beam, which in turn was carried by 4 × 4 stub posts 6 in. high. A wire mesh screen enclosed this 6-in. height.

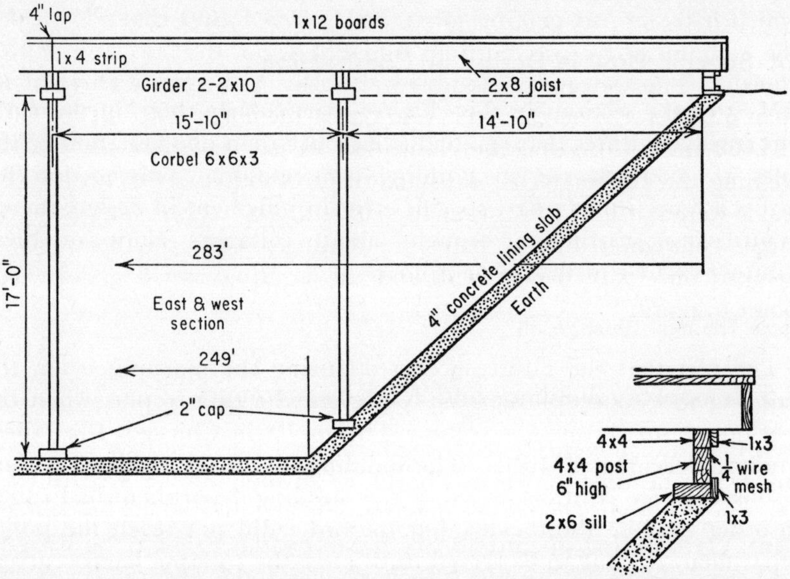

Fig. 7.20. Support of wooden roof of reservoir

Apparently the failure occurred when the sill tipped off the wall and slid down the 1:1 slope of the reservoir side, tipping the pipe posts and completely wrecking the roof. Figure 7.20 shows the complete lack of lateral bracing in the structure. The lack of anchorage of the sill made it readily susceptible to the movement that would be caused by the natural shrinkage of the wood members of the roof through the years.

Atlanta Auditorium—Atlanta, Ga.

On August 22, 1937, a section of the newly completed steel-joist roof of the Atlanta Auditorium [113] in Atlanta, Ga., collapsed just a few days before the newly rebuilt building was to have been dedicated.

The construction of the roof was a WPA project, and the design was in accordance with standard practice. The failure occurred during a heavy rainstorm, but investigation showed that faulty open-web steel joists, rather than overload, were responsible for the failure. The report of the investigating commission found that

Accumulation of water on the collapsed portion of the roof was not sufficient to have caused failure of a proper design. Your commission is of the opinion that the cause of the collapse was the failure of the open web steel joists supporting the roof deck.

The joists had been purchased from a company in Wheeling, W. Va., which had stopped making joists, and it was found that the steel in the web members was entirely different from the top and bottom chords. The web members tested all failed at a tensile stress of less than 60,000 psi, whereas steel from chord members failed at 90,000 to 125,000 psi. Moreover, the elongation varied, and tests showed that welding the two materials satisfactorily was extremely difficult. The welds were not made by automatic electric pressure welding, as by the larger companies, but by hand arc-welding. Tests on the welds developed only 450 lb per weld, and the welding varied greatly from point to point.

7.5 Temporary Buildings and Work by Amateurs

It is a strange coincidence that all three of the buildings written up under this heading are church buildings of some kind. Regardless of the use to which a structure may be put, certain axioms are self evident: (1) building operations are hazardous and should be designed and carried out by people who have had some experience in such things; (2) money saved on free services is often extremely costly; and (3) even a temporary building is subject to the same type of stresses as a permanent structure. Only very sketchy accounts of any of these failures were obtainable, but the information is sufficient to impress the above statements on anybody's mind.

Temporary church building—Barberton, Ohio

The hazards of construction are emphasized by the failure during erection of a temporary building [114] to be used for revival services at Barberton, Ohio, December 26, 1910. The actual cause of the failure was not made public. The construction was being done by volunteer workers of Barberton, and about 50 amateur carpenters fell 30 ft with the roof. One man was killed and 20 others were injured.

Peoples' Tabernacle auxiliary house—New York, N.Y.

On December 30, 1905, a reinforced-concrete floor [115] failed during construction without causing serious injury to anyone. The building was a six-story building being erected as a church auxiliary building for Peoples' Tabernacle, New York, N.Y. It had been designed by the pastor and was being built by day labor under his supervision. The floor was a combination tile-and-joist floor. The concrete was being poured for the fifth floor, which was shored from the fourth floor. Some of the shores were in place under the fourth floor, which had been poured two weeks earlier, and when these were removed an

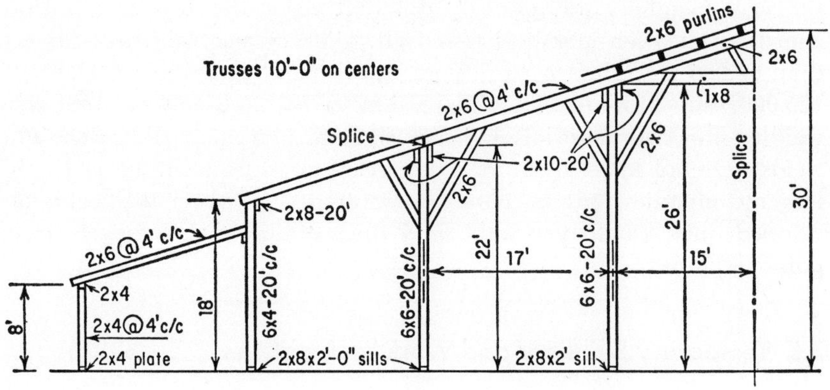

FIG. 7.21. Original framing of meeting house

area about 20 × 24 ft of the fourth and fifth floors fell. The third floor held, however, without failure. It is possible that there were contributing causes, such as the operation of an adjacent hoist and some concrete of questionable character.

Temporary timber meeting house—Lawrence, Mass.

A temporary wood building [116] to be used as a meeting house in Lawrence, Mass., suffered a partial collapse, while it was still under construction.

The total area of the building was 132 × 240 ft in plan, in which 30 × 240 ft failed. The lower chord of one 30-ft truss broke under a heavy snow load and, because of lack of proper lateral bracing, the side thrust caused the balance of the trusses to slip off their bearings, thus extending the failure the whole length of the building (Fig. 7.21). In the reconstruction the trusses were strengthened and additional 2 × 8 cross ties were placed (Fig. 7.22).

Comment in the engineering press following this failure emphasized the danger of building temporary buildings with less care and safety precaution than would be required for a permanent building. Because of wartime expediency, a number of buildings were built in industrial plants as "temporary buildings," during World War I and II. The author knows of buildings built in 1918, "for the duration of the war only," that are still in use in one chemical plant. Similarly, several such buildings built between 1941 and 1945 for a "maximum 5 years' life" are now approaching voting age and are still in use.

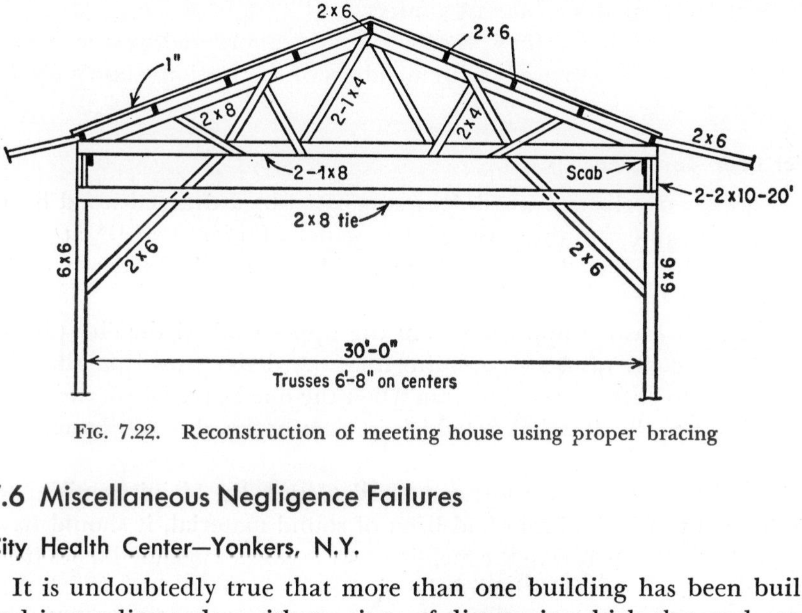

FIG. 7.22. Reconstruction of meeting house using proper bracing

7.6 Miscellaneous Negligence Failures

City Health Center—Yonkers, N.Y.

It is undoubtedly true that more than one building has been built and is standing today with no signs of distress in which the steel contractor has deliberately used steel of a lighter section than required on the architect's drawings. Throughout the entire range of rolled steel beams, three or more groups of beam sizes can only be told apart by micrometer measurement and frequently the specifications do not provide for testing laboratory inspection to make such identification.

In the case of the City Health Center building [117] in Yonkers, N.Y., in 1931, the attempt on the part of the steel contractor to use lighter sections proved to be a costly gamble. At this time American Institute of Steel Construction code called for an 18,000-psi fiber stress in structural steel, but the Yonkers code still used 16,000 psi, and the architect's design had been made on the basis of the latter stress.

Four stories of steel had been erected from a redesign by an engineer for the contractor before any question of adequacy was raised;

before a decision was reached on the disposition of the changes, the steel for the entire six floors was in place. An examination and analysis ordered by the mayor, disclosed the fact that some of the beams would not even qualify under the 18,000-psi stress. As for the variations from the sections called for, it was found that 55 out of 70 in the bottom tier, 43 out of 69 in the second tier, and 11 out of 63 in the top tier of columns were of a lighter section than was called for, and that 159 beams had been reduced in section. In addition, much of the riveting was poor, shelf angles had been omitted from spandrel beams, 28 of the columns had poor bearing and poor splice plates, and the construction was otherwise faulty.

As a result of the findings, the contractor was ordered to raze the entire frame, and to replace it with members of the sections shown on the original plans.

Pier floor—Long Beach, Calif.

A greatly overloaded wood floor supported by a decayed wood beam was responsible for the failure of a panel of pier floor [118] at Long Beach, Calif., May 24, 1913. The accident caused the death of 38 persons.

The floor was the approach from the upper deck of the city pier to the second floor of the city auditorium. A dense crowd on the pier was about to enter the auditorium when the one beam failed, dropping them 13 ft to the floor below, which in turn broke, dropping them 8 ft more.

The girder that broke was composed of two 2 × 14 joists spiked together. Even if the beam had been of sound material, it should have been 12 × 14. As it was, the member was decayed for some inches from its support. The condition of the girder was concealed by a scab that tied the girder over the top of the post to the girder in the adjacent span. The floor above, exposed to the weather, had no drainage and, in all probability, the rain water was retained in this area.

Bement-Darling Warehouse—Minneapolis, Minn.

The fact that any structure is only as strong as its weakest essential detail is illustrated by the failure, November 7, 1902, of a section of the Bement-Darling Warehouse [119] in Minneapolis, Minn., a six-story warehouse for farm equipment, which was less than a year old. No casualties or injuries occurred as a result of the failure, since the possibility of collapse had been indicated several days before it actually occurred.

The failure was investigated by Mr. C. A. P. Turner, a nationally known consulting engineer of Minneapolis. The story of the failure and its causes are extracted from the facts he unearthed and reported.

The building had $5\frac{1}{2}$-in. plank floors spanning between 12 × 20 wood beams, 7 ft on centers as shown in the plan in Fig. 7.23. These beams, approximately 21 ft long, were carried at one end by steel hangers set into the brick wall and at the other end by a row of wood beams and columns. The main units of the construction were of good, sound fir and were all capable of carrying the full design load of 250 psf.

An area about 22 × 60 ft on the sixth floor, given over to the storage of groceries, gave way. This area on the sixth floor was loaded to not more than 220 psf, the heaviest load of any section within the building. How long this load had been on the floor was not mentioned in the accounts of the accident, but several days before the failure a weakened condition was indicated by what the *Minneapolis Tribune* termed the "bulging of some of the upper floors," probably a sagging or deflection. After examining the building on the day before the failure, the architect stated that, in his opinion, it was safe.

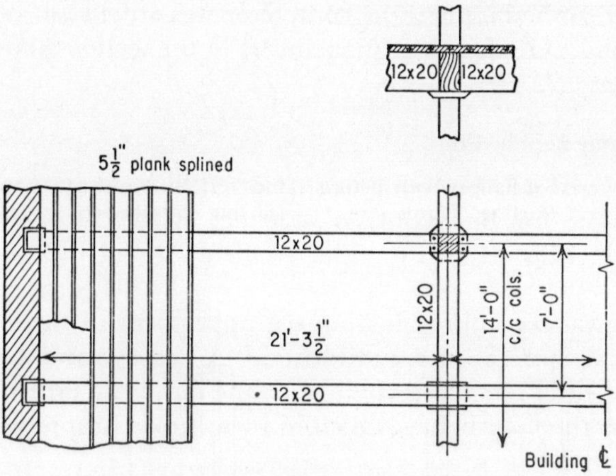

FIG. 7.23. Floor construction used in Bement-Darling Warehouse

On the morning of the accident, two of the officers of the company went to the fifth floor for further inspection and, while they were awaiting the arrival of the architect and contractor, the failure occurred—the failure of the sixth floor—carrying all the floors below it in this area through to the basement.

The point of failure was not in the members themselves, but rather in the beam hangers by which the ends of the beams were carried into the wall and to the girders and columns. The method of construction of the hangers is shown in Fig. 7.24 through 7.26, and the deformation and failure of the hangers which precipitated the failure is shown in

Fig. 7.27. The hangers for the wall and beam end varied slightly but, in general, they were ¼-in. thick plate, bent and welded, or riveted, as shown. The wall-bearing plate was 8 × 16 in., riveted to the 2-in. flange of the hanger body by four ⅜-in. rivets.

Aside from the elements of possible failure induced by the sharp bending of the legs of the hangers—probably by cold bending—the hangers, even under perfect conditions of manufacture, were stressed far beyond the ultimate strength of the steel. The surprising factor

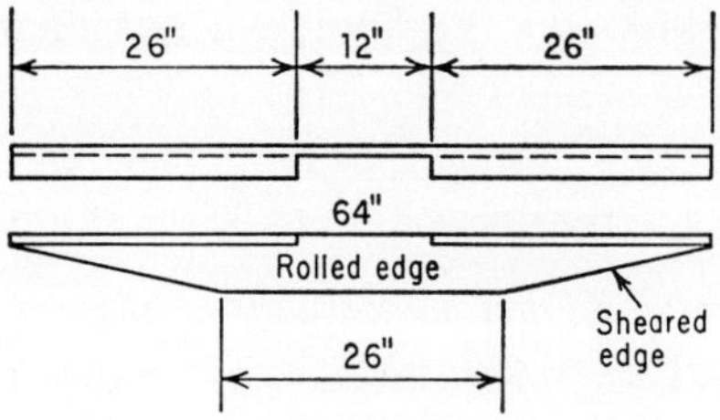

FIG. 7.24. Material of hanger body before bending—Bement-Darling Warehouse

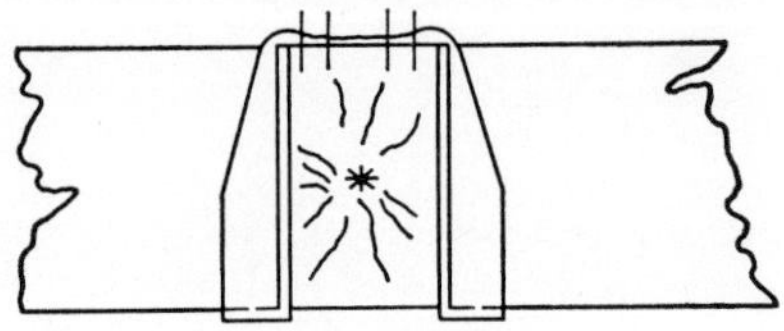

FIG. 7.25. Finished hanger—Bement-Darling Warehouse

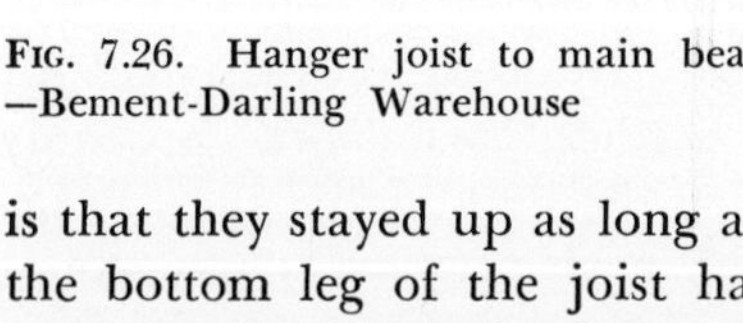

FIG. 7.26. Hanger joist to main beams—Bement-Darling Warehouse

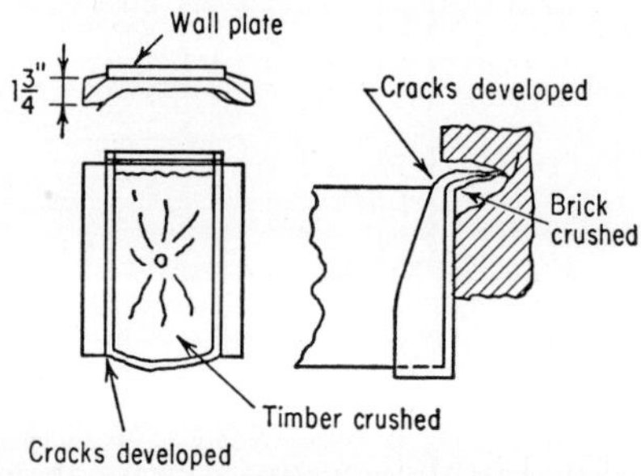

FIG. 7.27. Wall hanger deformed under load—Bement-Darling Warehouse

is that they stayed up as long as they did without failure. As it was, the bottom leg of the joist hanger at the wall buckled downward, crushing the side grain of the wood beam. At the same time the brick crushed under the concentrated loading of the upper leg of the hanger as it failed and bent.

Similarly, at the inside end of the beams, even before the failure of the hanger, the side grain of the main girder under the hook end of the hanger was stressed far in excess of the crushing capacity of the timber. The hangers in the lower floors failed in a different manner than those above, probably because the impact type of loading and the lighter loads they carried before the failure, prevented their being deformed as shown in Fig. 7.27.

Had the joist hangers been designed properly to distribute their load without overstressing either steel or timber, this failure would not have occurred. It is well to note, however, that this method of carrying load into a masonry wall by means of hangers instead of by direct bearing into the wall is not good practice and is not permitted by most building departments.

8

Lack of Precedent

Although the statement "It could happen to any of us," would apply to a large proportion of the failures in this book, it would be more likely to happen to those that occur because there is lack of sufficient precedent to enable us to know all we should know about a material or system of construction. No new method of putting materials together to form a building comes into being complete and with all the kinks worked out of it. The original designer may have conceived a method which, properly executed, would have been a satisfactory and safe structure, but he may not have had full control over its construction. Or he may have succeeded in having his brain child built as he had intended that it should be—and his structure is hailed as a forward step in engineering. Perhaps several such structures are built, and then someone who does not have the complete know-how undertakes a similar structure but fails to grasp some essential point in the original design. Often the point of weakness is in some detail that in ordinary standard construction is simply taken for granted, but that in the new design is subjected to additional or different stresses from those normally encountered.

Undoubtedly our successors in the next generation of architects, engineers, and contractors will consider failures of the type listed here as being the ordinary type of failure when the rules of the game are not followed.

8.1 Rigid Frame Buildings

These two failures are described together because they are similar failures of a standard design, which had been used with slight modification in 20 locations.

Wilkins Air Force Base warehouse—Shelby, Ohio, and Robins Air Force Base warehouse—Macon, Ga.

The first failure, at Wilkins Air Force Base, Shelby, Ohio, occurred August 17, 1955, and the second, at Robins Air Force Base, Macon, Ga., occurred September 5, 1956.[120] Both buildings had been completed and used for a relatively short time. These two failures occurring so shortly after the Coronado Naval Base failure described in the following article, focused a great deal of attention on this type of structure and prompted immediate changes in the ACI code for reinforced concrete.

The Wilkins Air Force Base warehouse was built from a design calling for reinforced-concrete beams 33 ft long poured monolithically with continuous, rigid frames, carrying poured-gypsum slabs. The frames were in a 67-ft span of a continuous series 400 ft long, with an expansion joint near the midsection.

The general contractor proposed a variation in the interest of economy, which allowed the substitution of a joist system using $8 \times 8\frac{1}{2} \times 17$-in. deep machine-made blocks tied together with bridge wire strands. Negative reinforcement was provided by stressing the wires in slots in the forms, and the joists and frames were poured monolithically.

The area that collapsed was about 4,800 sq ft and consisted of one 67-ft frame, which pulled down three more bents as it failed (Fig. 8.1 and 8.2).

Following this failure, inspection of several other warehouses from the same general design indicated that serious cracking occurred during construction or in periods up to two years later in areas where reversal of stresses might occur—near points of inflection. The upper end of the cracks seemed to occur near the points where the negative moment steel over the columns was substantially reduced, or dropped entirely.

Generally, the cracks followed diagonal lines at angles steeper than 45 deg and, in some cases, started near the middepth and ran either up or down from there. These cracks occurred under widely varying construction conditions, as far as aggregate and other concrete and construction factors were concerned, and the investigators quickly reached the conclusion that "the failures and cracking cannot be attributed to errors in construction." This preliminary report also included the statement that "the major contributing cause for what appears an inadequate design is therefore attributed to the deficiency of current practice, and the provisions of the government building code with respect to the required areas of web reinforcing and extension of longitudinal reinforcement."

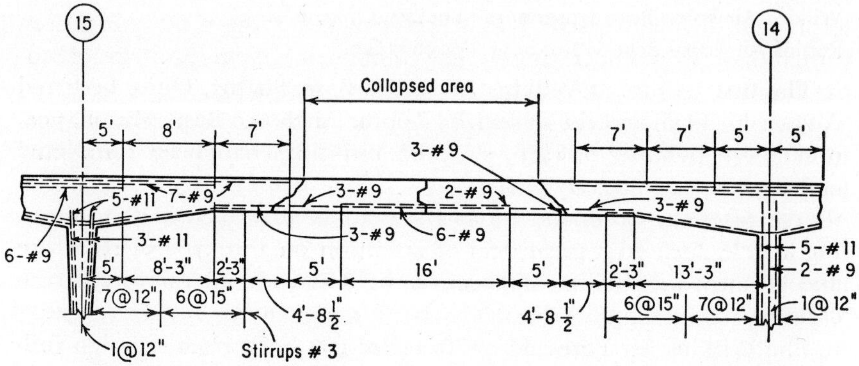

FIG. 8.1. Original design of faulty bent—Wilkins Air Force Base Warehouse

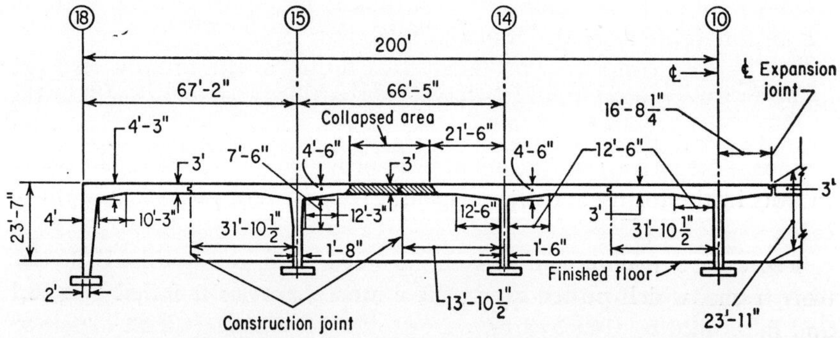

FIG. 8.2. Concrete frame at Wilkins Air Force Base warehouse

The final report by Amman and Whitney, consulting engineers to the Corps of Engineers, is summarized in the *Engineering News-Record* of January 12, 1956. In addition to repeating the statements quoted above from the preliminary report, this final report pointed to the fact that the overstress, which caused the failure, could only occur if the design and detailing did not anticipate either the magnitude or distribution of the actual stresses; i.e., if, for unseen reasons, either the stresses developed were greater than anticipated, or the resistance of the material was less. In all probability, the failure was due to a combination of several contributing causes. Neither the materials used nor the method of construction were at fault, and the plans would have been approved by any building department.

A high ratio of dead to live load (4:1) and consequent lower safety factor against working load would contribute toward the truth of the above statement. Moreover, it was found that in many instances the stress was built up because the expansion joint failed to work. In some instances, the steel plates that made up the joint were tack-welded to prevent their movement. In others, concrete around the joints prevented their proper operation so that volumetric changes contributed to the failure.

The time element was a contributing factor to this apparent change in stresses: creep, shrinkage, and reduced strength under long-time loading were partially responsible.

As a result of this report, the ACI immediately gave consideration to revisions looking toward heavier web reinforcement and extension of negative steel well beyond the point of contraflexure.

The recommendation was made that more than 300 frames in existing buildings be strengthened by the addition of external stirrups and that frames to be built in the future using this design should incorporate the following changes:

1. Web reinforcement to carry the entire shear in the region of low moment without help from the concrete.
2. Self-lubricating bronze and stainless steel rocker and expansion plates for the bearings in expansion joints.
3. Additional longitudinal steel in critical areas to compensate for effects of non-uniform live load, temperature and shrinkage.
4. Two continuous No. 5 bars in each vertical face to control local cracking.

Before the smoke of the Wilkins Air Force Base collapse had cleared away, a similar failure occurred at Robins Air Force Base, near Macon, Ga., in a building that had been in use for 18 months. The structure was similar to the one at Wilkins, except that the decking was of a different type, consisting of precast channel-type slabs resting on cast-in-place concrete purlins. The portion that collapsed at Macon was an area of 3,000 sq ft out of 64,000 sq ft total area. This building also utilized rigid frames of 67-ft span in a total series of 400-ft length.

Naval amphibious base—Coronado, Calif.

This failure of a single, rigid frame in the Naval Amphibious Base [121] at Coronado, Calif., occurred suddenly and without warning at 10:30 P.M. on February 6, 1954. There were no eyewitnesses, and an investigation was immediately begun to determine the cause of the collapse.

The mess hall portion of the building, where the failure occurred, was a rigid frame structure with precast reinforced-concrete bents and precast roof. The bents had a clear span of 67½ ft, spaced 20 ft on centers. The general dimensions of the bents are shown in Fig. 8.3. Investigation indicated that failure occurred in tension from the top corner toward the reentrant angle of the frame. Examination of the details indicated that a tensile splice existed at the point of failure, so that, practically, the bent was weak in tensile resistance at this point.

In the reconstruction, all the improperly detailed bents were torn down and replaced with bents having continuous reinforcement at the corners, with additional shear reinforcement around the corner and for some distance out on the frame (Fig. 8.4).

The failure brought about a great deal of discussion in the technical press concerning proper bar detailing and splicing for such structures, and the effect of proper bond.

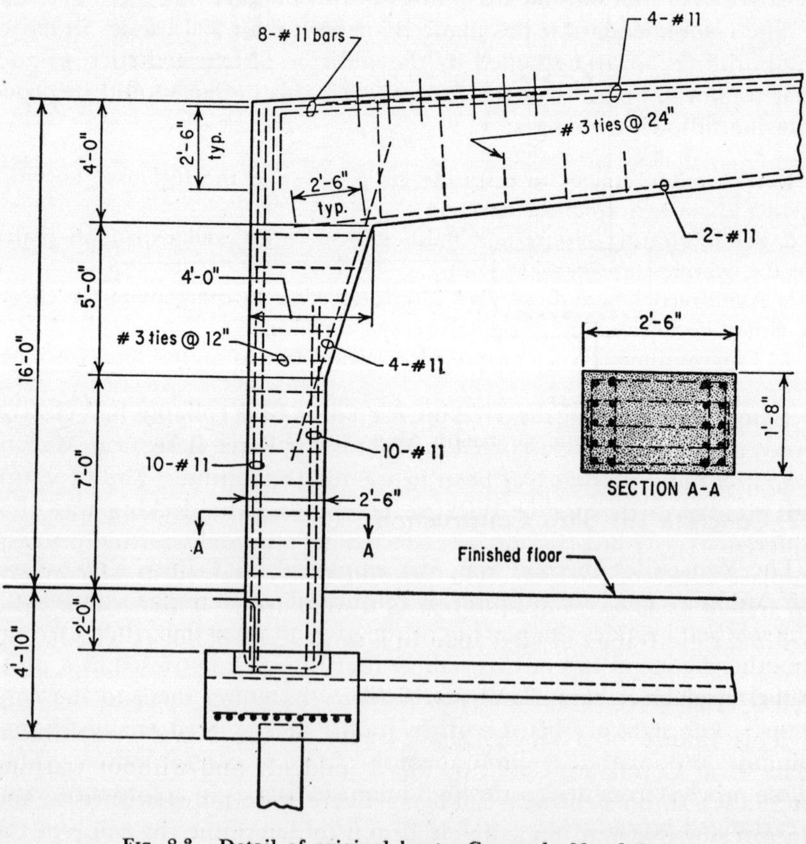

FIG. 8.3. Detail of original bents—Coronado Naval Base

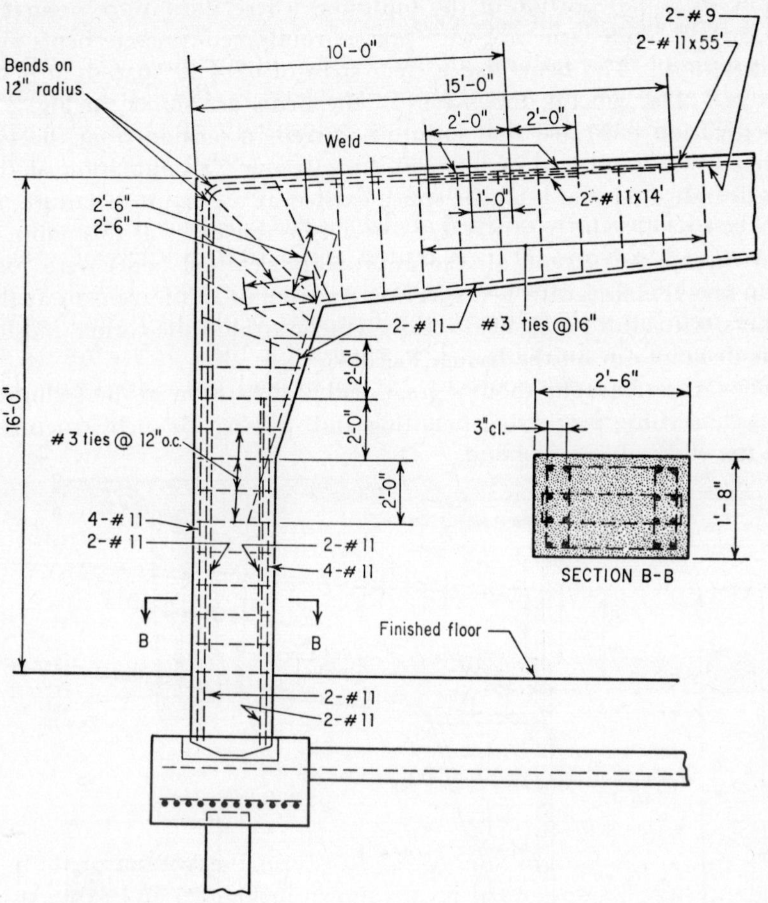

FIG. 8.4. Detail of rebuilt bents—Coronado Naval Base

8.2 Concrete Lift Slab Construction

The Youtz-Slick method was first employed at Trinity University, San Antonio, Tex., for erection of reinforced-concrete flat slab floors. The method assumes the placing of the ground floor slab, then casting the other floors on top of this slab so that they can be raised as a unit to their proper elevation and permanently fastening them to the columns. The slabs are lifted with hydraulic jacks placed on top of the columns and operated simultaneously from a central point. The whole process, from design through final welding, is in accordance with prearranged procedures.

Atlas Terminal Corp. lift slab—Ojus, Fla.

A series of three failures, all under different circumstances, attracted the attention of the construction industry to the Atlas Terminal Corp.[122] one-story lift slab building being built at Ojus, Fla., eight miles from Miami. The building was 100 × 200 ft with 14-ft ceiling height—the first of a large terminal of 40 units.

The roof structure consisted of eight waffle-type roof slabs, each 50 ft square with columns 30 ft on centers and a 10-ft cantilever overhang. The columns were, therefore, spaced alternately—30 ft and 20 ft, and there were no wall columns, the first row being 10 ft in from the perimeter of the building. The squares for the construction of the waffles were of precast concrete, 2 ft square and 13 in. deep with 2-in.-

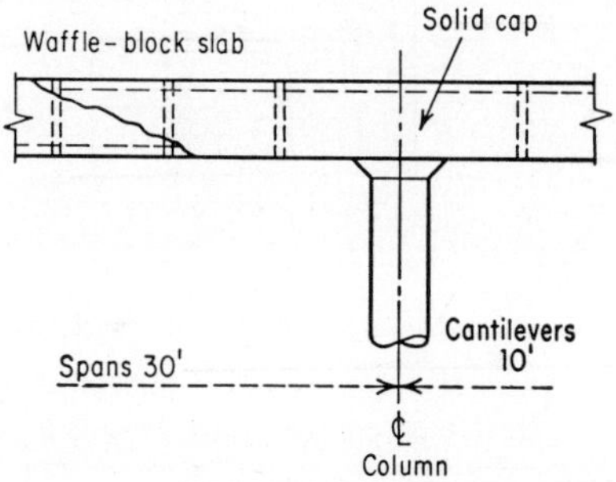

Fig. 8.5. Waffle construction—Atlas Terminal Corp. Building

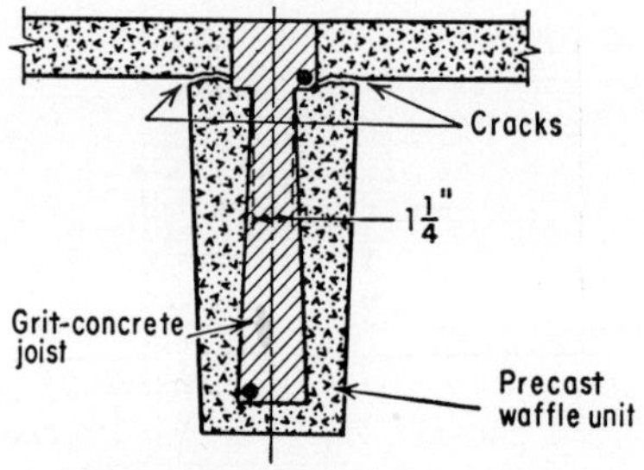

Fig. 8.6. Joist section—Atlas Terminal Corp. Building

wide lips at the bottom and all sides to form the bottom of the joist section. The section of the joist is shown in Fig. 8.5 and 8.6. At the column, one concrete block in each direction was omitted so that the slab at this point was a solid slab about 4 ft 4 in. square and 13 in. thick. The waffle blocks, which weigh 130 lb each, were cast of 5,000-lb concrete. No prestressing was used, and the construction was based on standard design methods.

The columns were 10 in. in diameter and were cast in place in cardboard forms. A metal pipe collar at the top of each column provided a means of attaching the roof slab. Aside from the use of a waffle slab, the method was the same as the Youtz-Slick method, and has been used in a number of other structures.

On this job, three of the eight slabs had been raised to final position and welded in place satisfactorily. The fourth slab was being lifted and it had been raised about 3 ft when, on March 18, 1952, it was noticed that one jack was lifting eccentrically. The jacks were supported on top of the columns by 5-in. pipe stub columns, and while the eccentricity was being corrected, the jack became unseated and dropped, breaking the slab and injuring the man who was making the adjustment. The slab was recast, and work proceeded on the remainder of the building.

On April 2, 1952, a second section fell, after it had been raised into place, injuring one man. The accident was attributed to stripping of threads on pipe stubs used for anchoring the slab to the column.

As a result of the failures, the county authorities ordered a load test of the roof before issuance of a certificate of occupancy. A load test was made, May 9, 1952, by flooding one panel with 8¾ in. of water, which represented full live load; both the maximum deflection and recovery were satisfactory. It was then decided to test another slab with a 50-per cent overload. When the depth had reached 12½ in. of the required 13½ in. of water, the slab failed and crashed to the ground. The failure gave warning before the crash, and nobody was injured.

Elwyn E. Seelye, Consulting Engineer, asked to make an inspection and report following this final failure reported that the cause was excessive shear on a section without stirrups. In some places the positive and negative steel were placed with so short a lap that the line of diagonal tension had no steel across its path. Moreover, the waffle blocks were so smooth that the bond between the poured concrete joists and the precast waffles was insufficient to cause the precast members to assist in shear resistance. The third finding was that in the waffle block itself, the joint between the horizontal top plate of the back and the sides of the box was so weak that cracks formed at this point. He called attention to the low factor of safety of the slab, but continued,

> We cannot say definitely, however, that the lifting operation might not have been the actual occasion of the failure of the first slab, when a column stub carrying a lifting jack was wrenched loose, or that defective threads on the inverted flanges that supported the slab were not the occasion of the second failure.

Pigeonhole lift slab garage—Cleveland, Ohio

The reader who is familiar with the pigeonhole parking garage knows that it consists of an elevator or crane operating down the center of the building by means of which cars are raised and parked on platforms both sides of the center well. Such a garage [123] in Cleveland, Ohio, had a floor system of two sections, each 21 × 91 ft in plan, each section supported by ten 61-ft high columns made up of two 6 × 6 angles welded toe to toe to form a hollow square section. The slabs were 8 in. thick, and each section weighed 90 tons. The garage was eight tiers high.

The procedure, as in any lift slab, was to cast the slabs, one over the other on the ground floor, then raise them to place, and weld them to the steel columns by means of a cast-in-steel collar. All slabs had been raised to their final position and wedged in place, April 6, 1956, ready for final welding. Early in the morning, winds with gusts of 35 to 50 mph hit the area, and on the morning of April 7, it was found that the frame had tilted to 7 ft out of plumb. The structure was braced until it was determined what measures to take. Upon inspection, the building inspector ordered the building dismantled. Instead, however, he permitted the contractor to jack and pull the structure back to plumb (Fig. 8.7).

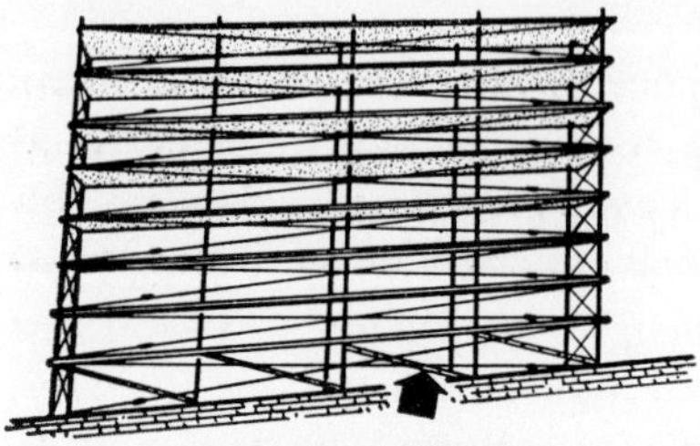

FIG. 8.7. Jacking tilted pigeon hole lift slab to plumb

Serra High School—San Mateo, Calif.

Although eight workmen and four spectators were injured when a lift slab fell at Serra High School,[124] San Mateo, Calif., while it was being raised, July 15, 1954, it might have been far more serious had it happened a short time earlier when about 60 spectators were standing on top of the slab.

The 8½-in.-thick slab, 65 × 70 ft in area, was to be supported on nine standard 6-in. pipe columns, each of which was anchored to the footing by four anchor bolts—½-in. and ¾-in. size. The columns were 16 ft high, unbraced except for the anchor bolts. It was the first lift slab in the San Francisco Bay area and invitations had been extended to architects and engineers of the area to witness the raising of the slab. About 60 of the spectators were on the slab when it had nearly reached its full height. It was noticed that the columns were 3 in. out of plumb, and the spectators were immediately cleared from the slab. In an attempt to pull the slab back into place with cable guys, the slab fell in the opposite direction from which it had leaned, landing 15 ft west of its original position and slightly to the south. Anchor bolts were sheared off or stripped in the collapse.

Following the failure, the engineer who originally devised the system placed the blame on the following four causes: (1) the design did not permit any lateral or eccentric loading without bracing during lifting; (2) the subcontractor had not braced in both directions; (3) the load of the group of spectators probably caused unequal and eccentric loading on the slab; and (4) the lateral shifting of the slab created an eccentric load that caused column failure.

8.3 Flat-slab Construction

Flat-slab construction uses no beams or girders, but employs a concrete slab, reinforced in either two or four directions. Formerly, all flat-slab construction was required to have splayed-head columns, but recently the splayed head has often been replaced by steel framework built into the thickness of the slab.

Prest-O-Lite Co. Building—Indianapolis, Ind.

The failure of the three-story Prest-O-Lite Building [125] in Indianapolis, Ind., December 6, 1911, is of particular interest because it marks the first recorded failure of a flat-slab-type building.

In the Prest-O-Lite Building all the concrete had been poured and cinders for forming the roof slopes to pitch the water to drains were piled on the roof. Of the 40 to 50 men who went down with the collapse, nine died and about 20 others were more or less seriously injured. The building was 70 × 88 ft in area.

The flat-slab system—now used quite generally for heavy duty construction such as warehouses—was being tried out in a number of different forms at the time; some of them were patented and others were free for anyone to use. There were no generally accepted formulas and their use was not even permitted in many cities. The slabs in this building were intended for live loads of 250 psf for the first floor and 150 psf for the second and third floors.

The shores were in place from the roof to the third floor and from the third floor to the second floor, but the shores for the second floor, which had been placed on October 23 and 24, were removed on November 15. The weather was reasonably warm for that time of year up to November 11, when the temperature dropped suddenly and stayed below normal until after the failure.

The collapse was blamed on several different possibilities: (1) possible lack of sufficient strength in the shores; (2) disturbance or inadvertent removal of one or several shores; (3) insufficient strength in the second floor slab; and (4) an overload caused by the pile of cinders on the roof. In the light of today's knowledge, it is possible that shear through the column cap or lack of sufficient negative steel at some point may have been the cause.

8.4 Delayed Failures

Gratiot School—St. Louis, Mo.

There are undoubtedly more buildings standing that should, according to all rules of engineering, have fallen down, than there are failures. Occasionally a delayed failure brings such a condition to light.

The Gratiot School in St. Louis, Mo., was built in the childhood days of reinforced concrete, when such things as diagonal tension and provision for negative moment were rather visionary. Built in 1898, it should have failed during construction, or at least under the snow load of the first winter. However, it stood safely for 17 years, and even then only one panel failed during a heavy wind and rainstorm, June 11, 1915, fortunately at a time when the building was not occupied.

The building was an interesting attempt, in the early days of this type of construction, to use the new material in conjunction with materials already tried and proven, such as brick walls and steel beams. It was a two-story-and-basement building with brick bearing walls 24 ft on centers. On these walls, spanning the classrooms, were steel I-beams 16 ft on centers, and between the beams and framed into the depth of the beam were 3 × 15-in. reinforced-concrete joists, spaced 18 in. on centers with a 2-in. unreinforced-concrete slab resting on the joists. It is not clear whether this slab was poured monolithically with the joists. The reinforcement of the joists consisted of one 1-in. square deformed straight bar to each joist, extending into the concrete fireproofing around the steel beams, with no reinforcement over the steel beam. The slab was apparently poured level, and roof drainage was provided by means of cinder concrete fill varying to a maximum depth of 9 in. at the high point. Despite the impact of 28 tons of concrete falling to the floor below, the second floor withstood the impact of the blow with comparatively little damage.

After the failure, examination of the remainder of the building showed shear cracks at the support of other joists and separation of the slab from the joist, particularly in the corresponding panel at the opposite end. Most of the ribs were badly deflected, and many of them had diagonal tension cracks.

The brick wall at the opposite end wall was about ¾ in. out of plumb, apparently pushed out by the expansion through the years of the cinder concrete roof fill. From this observation it was assumed that the same condition had probably existed at the end where failure had occurred and that the storm had merely provided the final nudge to bring about the failure.

Holland Building—Seattle, Wash.

It is interesting to note how methods considered proper in concrete construction of 1909 are looked upon in 1924 as being the cause of incipient failure. The Holland Building [126] in Seattle, Wash., a reinforced-concrete building, was built four stories high in 1909 with provision for four additional stories later.

The original design was in accordance with the best accepted practices of the day. The columns were spirally hooped with 1-per cent spirals and 2–2½ per cent of vertical steel. In order to reuse the octagonal column forms throughout, the columns were made the same size, varying the reinforcement and the amount of cover over the steel.

The building behaved satisfactorily until the additional four stories were added in 1923. When the construction of the extension was almost completed, serious cracking began to occur at the top of the interior columns. Shores were placed to relieve the columns of their load and the old concrete was investigated. The concrete in the columns had been made as a very wet mix, excessive agitation had caused so much segregation of the aggregate that the column tops were largely sand, the cement and coarse aggregate having settled to the bottom of the column. As a result, the tops of the columns were soft and porous with comparatively little strength to carry the load. In addition, a scum of laitance had risen to the top of the section and overflowed into the beam form—a layer which flaked off the bottom of the beams.

Good construction nowadays demands a concrete with as little mixing water as will make a workable mix, agitated with mechanical vibrators.

8.5 Prestressed Roof Construction

Blankner Elementary School—Orlando, Fla.

Despite the fact that properly designed buildings should not fail at all, there is at least a slight satisfaction in that many buildings with inherent weakness fail before they are occupied. The failure of a section of the roof of the Frances L. Blankner Elementary School [127] at Orlando, Fla., occurred during a rainstorm in the early morning hours of August 31, 1956, one week after the construction had been finished. Part of the classroom roof of the same building had been load-tested to 150 per cent of the design load and had been declared safe.

The roof system used was of a precast, prestressed double-tee system, which has been coming into use throughout the United States during the past few years (Fig. 8.8). Although it was designed for a live load of 20 psf, it is estimated that it failed under an average weight of 10 psf of ponded rain water. The area that failed was 32 ft in width of a cafeteria roof of 56-ft span, and the precast system was bearing on an 8-in. concrete masonry wall with a 7-ft-wide opening. Eight of the precast sections of the 56-ft length fell.

Investigations were instituted by the Prestressed Concrete Institute and by the Central Florida Chapter of the Florida Society of Engineers. The report of the Prestressed Concrete Institute stated that: (1) the collapse was not caused by ultimate failure of the slabs; (2) the design of the roof was defective because the extreme fiber stresses exceeded the allowable figure; (3) cracking of the slab resulted in excessive de-

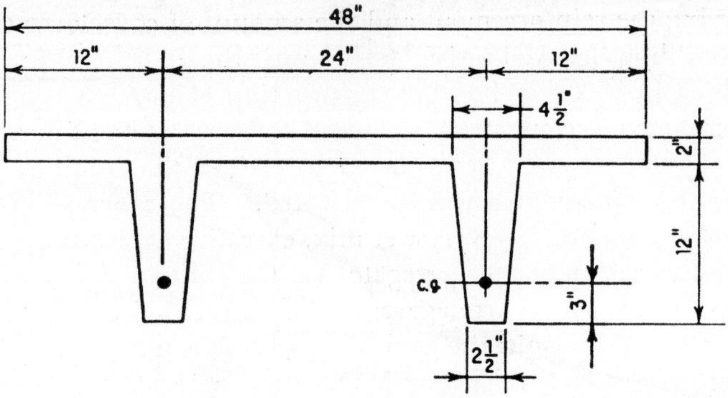

Fig. 8.8. Roof slab of Blankner School

flection and thus ponding of the water on the roof, which brought about more deflection and also rotation of the slab at the supports; (4) bearing points of the slab on the concrete block wall were eccentric, and the wall design was inadequate; (5) the rotation of the slabs shifted the bearing point, causing the wall to collapse; (6) the wall had window openings from one end to the other; and (7) the pilasters and the bonding beam were improperly designed. Their report included a graph (Fig. 8.9) showing the stresses developed and indicating that with a 10-psf live load, the concrete was carrying a tensile stress of about 800 psi.

Delving further into the details of their report, we can see from Fig. 8.8 that to prevent prestress from being transferred from the steel to the concrete, the bond between steel and concrete was broken by paper tubing through part of the length of some of the pretensioning wires. The tension stresses are thus transferred into the concrete and, the tension cracks in the concrete show up at the point of the maximum tensile stress in the concrete. The computed deflection is about 1.5 in.

The report of the Florida Society of Engineers points to a bearing failure, either at the base of the stem of the tee or in the bone beam—the bonding beam along the top of the 8-in. concrete masonry wall.

The fact that the slabs took a 50-per cent overload in a test on a classroom wing—a span of probably not more than 24 ft—is a strong indication that the deflection of a span of 56 ft was definitely the cause. The total depth of the construction, 14 in., was one-forty-eighth of the span and, under these circumstances, becomes a great factor in the

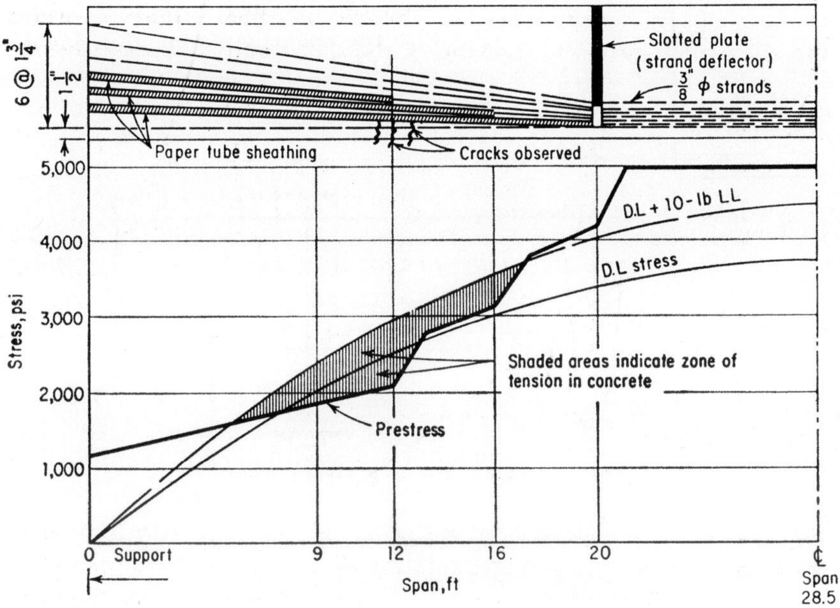

FIG. 8.9. Plot of estimated bottom fiber stresses—Blankner School

deflection of a beam and consequent transfer of the bearing load to a point bearing at the edge of the support, inducing a dangerous eccentricity.

When a building failure occurs without any clear evidence of the reason, disagreements frequently arise between the various parties involved, the architect, engineer, contractor and owner, as to the cause. These disagreements are compounded when the failure involves some material or method for which there is not sufficient precedent to point up some common usage.

Community Grape Corp.—Lodi, Calif.

In the case of failure of a new concrete warehouse [128] for the Community Grape Corp. of Lodi, Calif., in October, 1952, problems of prestressed structures and of Vierendeel Trusses posed two construction problems for which there was lack of sufficient precedent.

Both prestressed concrete and the theory of Vierendeel Trusses are difficult to describe to a layman. Usually, prestressing is employed with precast beams. The general theory of prestressing may best be indicated by comparison of Fig. 8.10 with Fig. 1.4. By using a steel capable of taking very high tensile strength, the steel is prestressed before the beam is placed, thus lowering the neutral axis almost to the plane of the steel and increasing the area of the compression triangle of stress. The Vierendeel Truss is a framework in which there are no diagonal members, so that instead of each truss member carrying direct tension or direct compression only, the members are dependent on the ability of each of the joints to resist the bending that exists in each of the members.

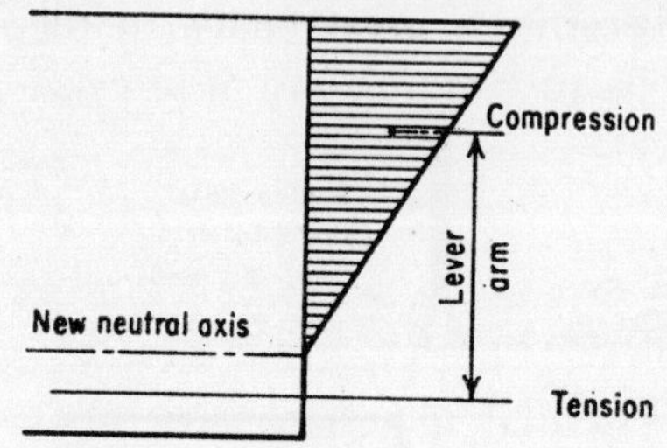

FIG. 8.10. Theory of prestressing (compare with Fig. 1.4.)

The trusses (Fig. 8.11) were 80 ft long, and the structure was about 90 per cent complete when it failed. The accident occurred during the night; therefore, there were no witnessese and nobody was injured.

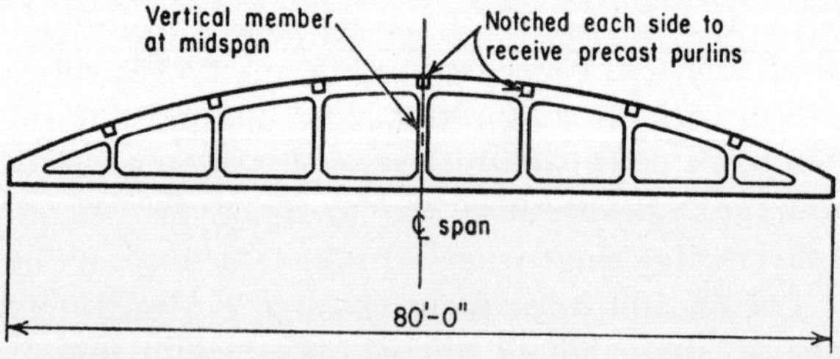

FIG. 8.11. Prestressed concrete Vierendeel roof trusses of Community Grape Corp. Warehouse

The failure was brought into the engineering press when the contractor filed a complaint of incompetence of the designer with the California State Board of Registration for Civil and Professional Engineers. The contractor claimed that difficulty had been encountered with the trusses earlier, and that additional wires had been placed on the lower chord outside the concrete. The fact was also cited that the truss had no center post. The designer countered with the statement that the difficulty that had been encountered earlier had resulted because the contractor had attempted to pick up a 19-ton truss with two small cranes. One crane had tipped over, dropping the truss on several other trusses, and that to meet a time schedule, the injured trusses had been reinforced by the additional wires described. With reference to the missing center post, he called attention to the fact that many of the bridge trusses in Belgium, where this type of truss had originated, were built without a center post. He also stated that the contractor had not followed specifications in cleaning the prestressed wires in order to get a better bond.

8.6 Miscellaneous Newer Systems

Balloon-formed structure—Los Angeles, Calif.

Wartime restrictions on the use of essential materials have sired some interesting methods of using both concrete and wood in lieu of steel.

One such system involved the use of an inflated canvas bag in place of wood forming for the placing of a concrete dome. The failure of such a dome occurred November 18, 1943, in a building [129] under construction in Los Angeles, Calif., which was intended to be a marking and sorting room for a linen service. The structure was to have been a reinforced-concrete dome 100 ft in diameter and 32 ft high at the peak.

The design was not unusual; the method of accomplishing the result was. Without a doubt, all of the troubles encountered were the result of lack of sufficient precedent. The method involved inflating a canvas bag form with air compressors, and maintaining the necessary air pressure while a 1 to 1¼-in.-thick coat of cement grout was gunited ("shot on" under air pressure from the nozzle of special equipment) in quarter-inch layers and rings from 6 to 10 ft deep, starting from the spring line. The full thickness for each ring was built up before proceeding to the next ring. Then, having completed a dome 1 to 1¼ in. thick, the required reinforcement was placed and the remaining layers of gunite were placed to build up the finished reinforced-concrete roof.

During construction of this, the first structure to be built using this new system, two failures occurred. When the first coat of gunite was being placed, it was found that the canvas balloon form was capable of carrying a pressure of only slightly more than ½ oz, instead of the necessary 1½ oz, and this failure caused a delay of two days while a reinforced-balloon form was being prepared and set up. The second failure occurred when one of the three blowers in use failed, and permitted the pressure to drop to ¼ oz per sq in. before another blower could be brought into action. The structure failed with a crash, but none of the 20 men carried down with the scaffold were seriously hurt. After repairs, the structure was completed.

Forum Building—Vancouver, B.C.

The lamella-type roof is a form of wood arch made up of short-length wood members formed together in a diamond-shaped lattice

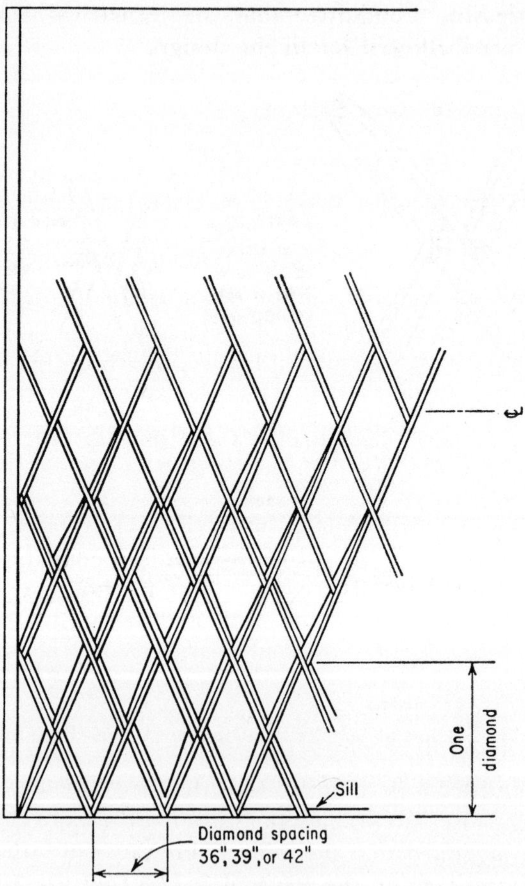

FIG. 8.12. Plan of lamella construction—Forum Building roof

pattern of joists as shown in Fig. 8.12–8.14. Note the shape of the lamella in Fig. 8.13. Curvature is obtained by cutting upper edge only. This kind of roof has all the good points of arch construction under uniform load with the added bonus of economy, but it also has the instability of arch construction under concentrated or one-sided load (Sec. 1.4).

The Forum Building [130] in Vancouver, B.C., with a roof of this type, 126 ft span and 375 ft long, failed, January 21, 1935, under a very heavy snow load, which drifted entirely to one side of the roof. The design live load was 40 psf, and it would have been satisfactory, under normal conditions for Vancouver. Snowfalls are ordinarily not heavy in Vancouver, and what snow does fall is usually wet and does not drift. The engineer for the Lamella Roof Syndicate of New York told the Vancouver Investigating Committee that unsymmetrical loading of this type had not been allowed for in the design.

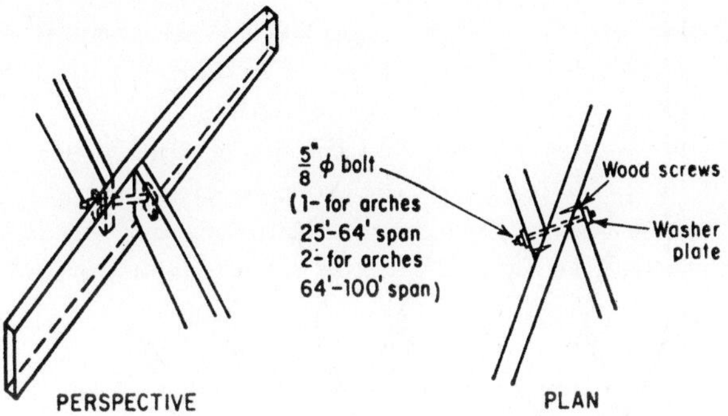

FIG. 8.13. Joint detail of Forum Building roof

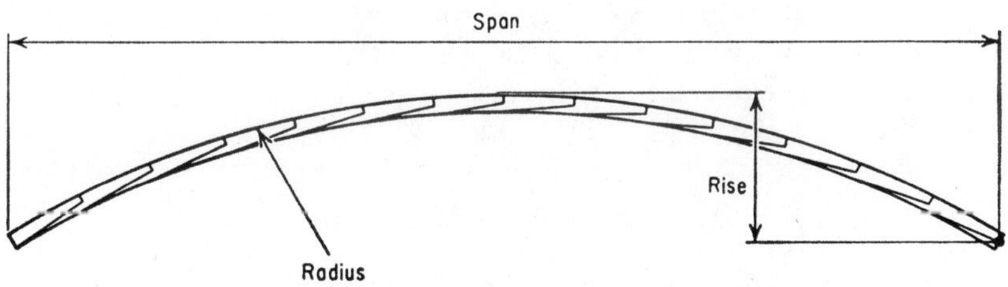

FIG. 8.14. Section of lamella construction—Forum Building roof

A later committee report agreed that the unsymmetrical loading was a contributing cause, but because the snow load was only approximately 15 psf, it was suggested that the more likely cause of failure was eccentricity of the total load on the columns. The roof sills supporting the lamella roof were carried by brackets at the column, producing an eccentricity of 1 ft 3 in. and this eccentricity rather than the eccentricity within the arch itself was probably the precipitating cause of the failure.

Union Carbide Co. Building—Toronto, Ont.

The Union Carbide Co. Building [131] was to have been 215 ft long by 65 ft wide and eleven stories high. It was a series of welded rigid frames 20 ft apart; each frame consisted of the two exterior columns and the floor beams across the building, which spanned the entire width of the building. Apparently, until the floors were in place, only the light steel spandrel beams furnished horizontal stiffness to the structure. Although each frame in itself was rigid laterally, there was very little stiffness at right angles to the frames. Temporary bracing of steel cables held the frames, and the welding was completed through the seventh floor.

The failure occurred about 6:30 P.M. on Saturday, September 6, 1958, when, under the impact of wind, the two upper stories started swaying, and the building fell longitudinally. No casualties occurred because of the time of the failure, but the collapse caused a property loss of $600,000 in the 1,850-ton steel structure.

Although investigators agreed that the direct cause of the failure was the wind, it is probable that the comparative slenderness of the horizontal bracing members was a contributing cause. It is interesting to note that in the reconstruction, the use of deep horizontal trusses was contemplated instead of the spandrel beams.

The failure was an unprecedented test of the welded joints in a structure of this kind because only one of several hundred welds failed, despite the fact that the stresses encountered must have been far in excess of design stresses. The members themselves were badly bent and twisted, but the welds held. A check revealed that the foundations showed no settlement, and the reinforced-concrete substructure came through in excellent condition.

9
Old Buildings and Overload

In August 1930, after the collapse and incipient collapse of a number of old buildings in New York City, there appeared in the *Engineering News-Record* a letter from David M. Oltash advocating regular inspection of old buildings. Usually, such a suggestion would not be kindly received by building owners as it would involve cost—the cost of an experienced inspector in the first place, and the possible cost of repairs if the structure is found wanting. The owner prefers to bury his head in the sand on the assumption that the building has stood all these years and there is no reason to question its safety now. He is a firm believer in the "good solid construction of the old days. They don't build like that nowadays."

9.1 Need for Periodic Inspection

And yet, the collapse of an old building emphasizes that, for a number of reasons, these old buildings should be periodically inspected. Eliminating the causes peculiar to New York City and a few other cities, such as subway or tunnel construction, the reasons listed why this inspection is advisable are:

1. Overloading or change of use of building

2. Dry rot in wooden pile foundations caused by lowering of the water table

3. Materials used for bricklaying or stonework changing through age

4. Bearing capacity of brick walls or stonework changing as the result of cement aging (if any used)

5. Placing heavy loads in the center of spans, thus overloading floors

6. Removing partitions that may have originally been designed for bearing

7. Alterations made without the advice of either architect or engineer

The reader can probably add to this list. From the observations of the author, the following may be added.

8. Excessive vibration caused by the enormous increase in heavy traffic on adjacent highways

9. Increased knowledge in action of stresses and stress capacity in materials

10. Natural deterioration of wood and steel in the structure itself

11. Damage to unoccupied buildings by trespassers or vandals

Although it is undoubtedly true that *periodic* inspection would not have prevented some of the collapses that are described in the following pages, it is reasonable to believe that many of them might have been prevented by proper forewarning—at least lives might have been saved or injuries prevented.

In the Wisconsin Bureau of Engineering, one of the functions of the industrial division was the periodic inspection and investigation of structural conditions in buildings used by the public; the quarterly report of this department just prior to an article in the *Engineering News-Record* in 1930 noted two buildings—one of them a theater—where the timber roof trusses were so badly rotted at the wall bearings as to require immediate repairs. There is no doubt that this periodic inspection by the Wisconsin Bureau of Engineering has prevented serious accidents.

9.2 Failures in Old Wood Construction

As a construction material, wood is unique. It is the only organic material used in construction and is, therefore, subject to many phenomena that are not applicable to its inorganic counterparts—steel, concrete, and masonry. None of these other materials will decay as wood does. Steel, of course, will rust out, which, in some ways, is similar to decay, but it is not the same. Wood will change its volume, both with the grain and across the grain, with changes in moisture content, and these shrinkage changes produce stresses such as are produced in no other structural material. Wood is a perfectly satisfactory material if it is properly prepared, properly put together, and properly protected and inspected. Each wood failure is not so much a failure of the material as it is a failure on the part of somebody to do one of the things listed above.

Robinson Opera House—Cincinnati, Ohio

In old buildings built before the era of steel or wrought-iron beams, one of the most fruitful causes of roof failures was the rotting out of the end of a wood roof truss. As already described concerning the failure of wood piles, the end grain of a timber subject to alternate wetting and drying is particularly vulnerable to decay. When a building is built, the roof is tight and the mortar joints are sound so that no moisture can get to the end grain. After some years, however, leaks may develop in the flashing, or mortar may become permeable so that the ends of the timber trusses are no longer protected against moisture and a dangerous condition may develop.

Usually such failures are local, but although local, they can be dangerous, such as the failure of a single beam or truss, which, through proximity to a leak, permits water to get to the timber. Such a failure occurred October 15, 1897, when a truss in the roof construction of the Robinson Opera House [132] at Cincinnati, Ohio, collapsed, resulting in the death of four persons and injury to 30 others.

The building had been built in 1870, using wood roof construction throughout, including timber trusses, which spanned 77 ft. During an evening performance, the decayed end of one of the central trusses sheared off, dropping the wood lath and plaster ceiling and a heavy chandelier on the audience below—together with the wrecked truss and ceiling joists. Apparently the roof itself behaved as a suspension bridge, for the purlins and rafters draped themselves between the trusses that remained and did not fail. If this further failure had occurred, undoubtedly the death toll would have been much higher. As it was, the ceiling which fell was from an area of 76 × 30 ft.

Wooden truss—Portland, Ore.

The necessity for properly seasoning the timber used for the construction of wooden trusses cannot be overemphasized. This fact was pointed up by the failure of a building [133] in Portland, Ore., July 23, 1927.

An old two-story brick building with brick walls and bearing partitions carrying wood floor construction had been altered nine years previously by removing several bearing partitions and substituting roof beams to carry the joists, which were suspended from new trusses placed above the roof. To protect the trusses from the weather, they had been enclosed in galvanized steel covering, thus preventing any possible inspection of their condition. The type of truss used—a Howe truss—is indicated in Fig. 9.1.

When further alterations were made in 1927, one of the trusses collapsed. As the drawing shows, the bottom chord was spliced in two

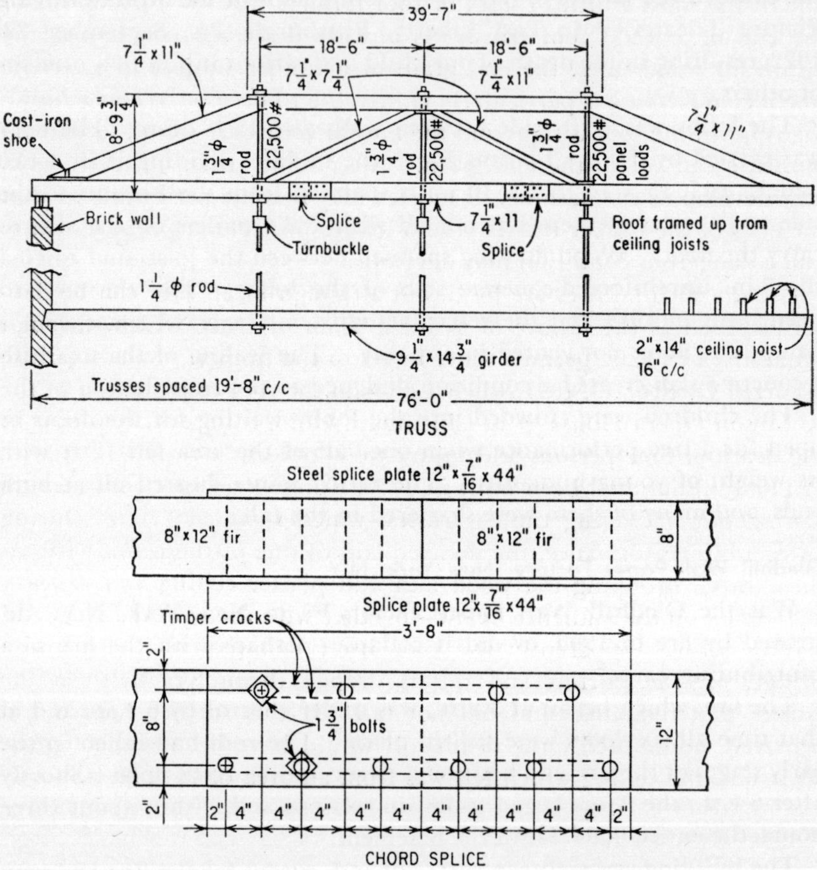

FIG. 9.1. Howe roof truss and tension splice of bottom chord

places, using steel splice plates. It was at one of these splices that the truss failed. Examination indicated that the wood failed when the splice had given way because of horizontal shear along the top and bottom of the bolt lines. When the rest of the trusses were examined, the same condition was disclosed in others. In six of nine splices examined, failure had started. Excessive lateral shrinkage was indicated—in all probability caused by the use of green lumber. The steel splice plates undoubtedly contributed to the failure because, although the timber shrunk laterally, the steel held the bolts in their original positions.

Strand Theater—Pittsburgh, Pa.

A condition ideal for the development of decay in wood joists was the cause of the failure of part of the lobby floor of the Strand Moving Picture Theater [134] in East Liberty, Pittsburgh, Pa., September 22, 1922, resulting in the death of one child and serious injury to a number of others.

The lobby was 32 ft wide between walls and 15 ft deep. The floor was carried by double I-beams across the middle, dividing it into two panels, each 15 × 16 ft, 2 × 10 joists spanned from the I-beams to the side walls, most of them but not all with 2 × 4 nailers on the side to carry the floor. Wood flooring spanned between the joists and carried the 5-in. unreinforced-concrete slab of the lobby. The theater had been built in 1914 and the cellar, actually only a crawl space, had a damp, dirt floor, not vented in any way. The sealing of the top with a concrete slab created a condition designed to promote decay.

The children were crowded into the lobby waiting for the doors to open for a free performance when one-half of the area fell 12 ft with its weight of young humanity. The rotted joints sheared off at both ends, and many of them were shattered in the fall.

Gledhill Wall Paper Factory—New York, N.Y.

Was the Gledhill Wall Paper Factory [135] in New York, N.Y., destroyed by fire in 1909, or did it collapse—perhaps with the fire as a contributing cause?

The fire, which began at 3 P.M., was under control by 6 P.M., and at that time all six floors were still in place. The roof had fallen in the early stages of the fire and had come to rest on the sixth floor. Shortly after 6 P.M., the floors began to fall in sections and within about three hours, the entire mass was in the basement.

The building was eighteen years old and, except for the first bay next to the street, was of heavy timber construction. The floors, roof, and wainscot, however, were of light, highly flammable wood, and the contents were, of course, extremely flammable. Certainly timbers of this size should not have burned out so easily, and the manner of failure indicated that it was not the fire which caused their collapse.

Investigation disclosed that the oak posts of the building had dry rot to the extent of being in a dangerous condition and that the real failure occurred when this punky wood smoldered after the fire had burned itself out.

9.3 Old Masonry Building Failures

A brick wall is not a homogeneous structure (although it approaches homogeneity) because the brick and the mortar with which it is laid have different shrinkage and expansion coefficients under temperature changes. In old walls, the mortar was usually a lime mortar, which changed its chemical composition through the years and permitted the infiltration of moisture, which froze and caused further deterioration. When all the potential possibilities for destruction of the wall are considered—particularly in a northern climate—it is surprising how well brick masonry has proven itself through the centuries.

A large proportion of masonry failures can be attributed to lack of sufficient bracing or anchorage. The Majestic Theater failure is a typical failure of this type. Moreover, many of the old masonry buildings have undergone change after change—new openings cut in walls and old openings bricked up without proper bonding—changes on top of unrecorded changes. In addition to these prime causes of deterioration, the tremendous increase in heavy traffic has contributed to almost every masonry failure in old buildings.

All in all, as has been stated, brick masonry has had a satisfactory record and, within the limitations of modern building codes and modern materials of construction, today's buildings will have an even better record.

Empire Apartment Building—Washington, D.C.

The existence of ordinances and the issuance of instructions to do certain things do not in themselves insure the safety of a building. The orders must be followed through and, in some cases, without delay.

The Empire Apartment Building [136] in Washington, D.C., was a 54-year-old wall-bearing apartment house, seven stories and basement in height. The building was 96 × 146 ft in area, with bearing walls varying from 13 to 27 in. in thickness.

Early in the century a hole had been cut through one of the wall-bearing partitions in the basement, and in 1932, the opening had been enlarged. Later more changes were made from time to time—not all of them in compliance with code regulations. The result was the overloading of a brick pier in the first floor that had not been properly reinforced. The resultant collapse of a 30-ft square section of the building, December 16, 1947, caused the death of four and the injury of twelve other residents of the building.

Majestic Theater—Pittsburgh, Pa.

What might have been a major catastrophe was averted by the failure of a ceiling, May 15, 1922. Although people were injured, the number was far less than would have been hurt if the entire building had failed.

The Majestic Theater [137] occupied the rear two-thirds of an old church in Pittsburgh—an area of about 60 × 75 ft—the front third being used by stores. The theater was spanned by four timber trusses in the 60-ft direction, and the ceiling was of plaster on wood lath on 2 × 6 joists. These, in turn, were notched and carried by 1 × 3-in. ribbons on the sides of the bottom chords of the trusses. The joists were also

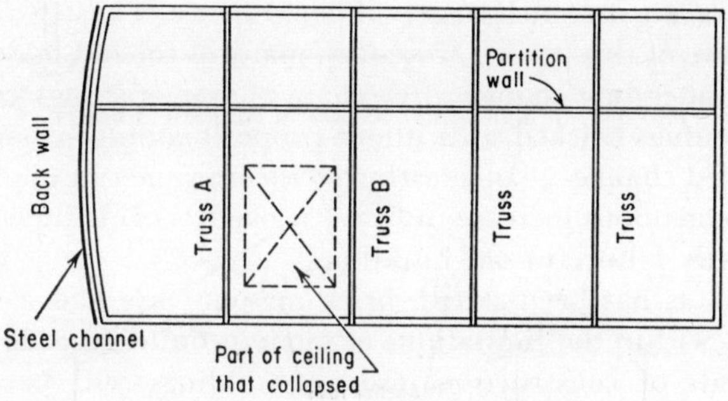

FIG. 9.2. Plan of Majestic Theater

toe-nailed into the truss chords. The walls carrying the trusses were 13-in. brick walls, 50 ft high, which had been laid in a very poor grade of lime mortar.

The end wall, however, was the one in the most critical condition. It was bowed almost 1 ft and, at an earlier date, an attempt had been made to hold it against further movement by placing a steel channel flatwise back of the wall, anchored at the ends and tied by means of rods to the first truss inside the wall. The wall had apparently been bowed before the plaster ceiling was installed, at the time of converting the church into a theater (see Fig. 9.2–9.4), because there was no separation between the wall and the ceiling.

Apparently the rods anchoring the channel pulled the first truss as the bow in the wall increased, and the distance between the first and second trusses increased accordingly, finally reaching the point where the ceiling joists slipped off the ribbons and dropped the ceiling on the people below. Fortunately, few people were at the theater that evening, and seats in the area where the ceiling fell were the least popular in the house. Most of the damage was done in an area of empty seats.

The possibility of a major catastrophe that did not happen lay in the condition of the brick walls. Walls, 13-in. thick, unsupported to a height of 50 ft are a hazard even with good masonry. When the mortar is such that it can be picked out easily with a knife, when some walls show disintegrated mortar, and when large cracks—which indicate by

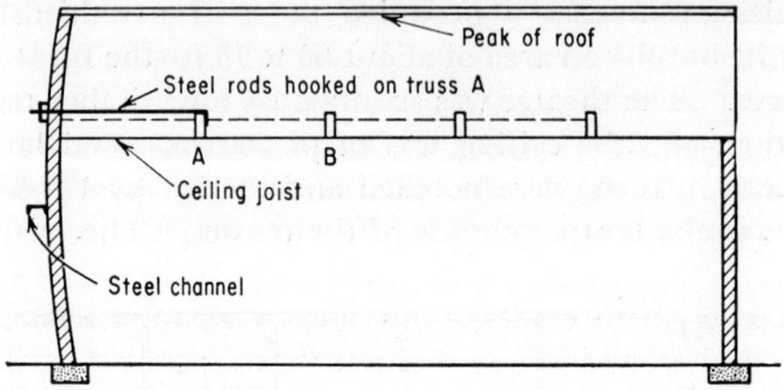

FIG. 9.3. Longitudinal section of Majestic Theater

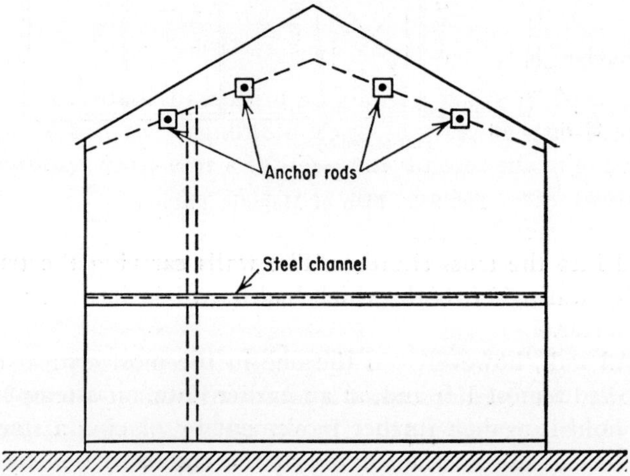

FIG. 9.4. Back and elevation of Majestic Theater

their discoloration that they are old cracks—are passed up by the building inspector, who only two weeks before the ceiling failure had pronounced the building safe, the suggestion of an incipient failure—to the extent of being a catastrophe—is certainly warranted.

Seven-story warehouse—New York, N.Y.

A typical failure of this type—without any signs of distress beforehand—is the collapse, June 20, 1930, of a 35-year-old warehouse [138] in New York, N.Y. This failure was described in the *Engineering News-*

Record of June 26, 1930, and from this article the following is extracted. The building collapsed, completely filling the site and the street in front of the building with a mixture of brick, broken timbers, and stored materials—mostly foodstuffs. Seven men, who were at work on the loading platform at the street level, escaped unhurt, and two men on the roof had minor injuries.

The building had brick walls and a timber floor system, with 4 × 12 beams carried by 12 × 14 girders. It was wallbearing on exterior walls and supported on interior columns, which were on spread footings of concrete on clay. The loading, although fairly heavy, did not appear to be excessive. It was at first believed that the failure started at the front wall near the top, perhaps through the settlement of interior columns.

There was very little evidence of either overload or serious deterioration of the timber, and it was thought that a possible cause may have been that the unanchored floor joists had pulled out of the brick walls progressively in the course of years—probably the result of traffic vibration.

Wall—Brooklyn, N.Y.

On August 12, 1914, a section of a brick wall about 8 × 12 ft fell out of the front of a 40-year-old brick building[139] in Brooklyn, N.Y. It was a section of the second story wall of a four-story building between two windows.

The failure was attributed to the excessive drying out of the mortar by the heat of an adjacent chimney, aggravated by the vibration from frequently passing elevated railway trains.

Marlborough-Blenheim Hotel—Atlantic City, N.J.

Normally, steel reinforcement is protected from corrosion by the concrete in which it is embedded. In the early days of concrete construction, however, the concrete was poured in the form of a very wet mix to insure that the steel was properly embedded and that the concrete would not honeycomb. This concrete mixed with an excessive amount of water—far above the amount required for the necessary chemical action—was apt to become porous as the excess water was drained out by evaporation. This, accompanied by the moist seashore air, was the combination that contributed to, and perhaps brought about, the failure of a small section of a concrete roof of the Marlborough-Blenheim Hotel[140] at Atlantic City, N.J., August 9, 1921 (see Fig. 9.5). An analysis of the design, however, indicated that the steel reinforcement had been greatly overstressed throughout the entire life of the building.

At the time it was built in 1906, there was considerable doubt as to the wisdom of building with reinforced concrete along the seashore without much greater protection of reinforcement than was normally given. Apparently the amount of cover varied in the area which fell, ranging from less than 1/4 to 1 in. The reinforcement was a welded wire mesh, and the collapse occurred by failure along a line near the center of the slab. The wires at the point of failure were slightly necked down and showed the typical cupped fracture of a tension break.

This failure had special interest for the engineering profession, because it was not only the first concrete building to fail through corro-

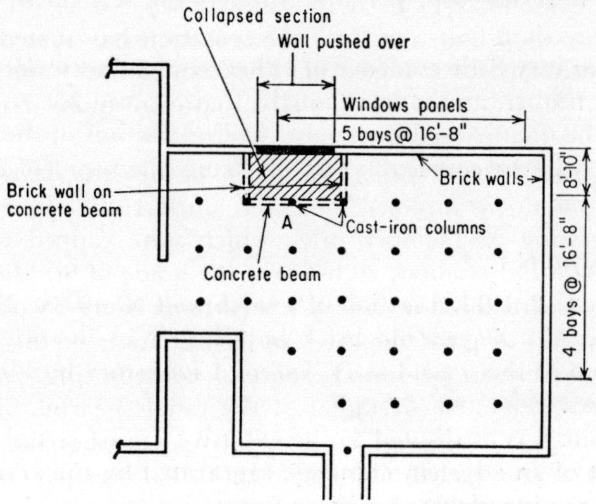

FIG. 9.5. Collapse in Marlborough-Blenheim Hotel kitchen annex

sion of the steel, it was probably the first reinforced-concrete building to fail after it had been completed and occupied.

The portion of the roof that fell was a section of roof slab over a two-story kitchen annex. It was of cinder concrete, which is a somewhat porous material and, if the wrong kind of cinders are used, possibly corrosive, even without the moist salt air. The slab was an 8-in. thick slab on a 15-ft span, covered with a bituminous coating over which was laid a cement finish. As has been noted above, under the dead load alone, the steel wire reinforcement was stressed theoretically to 100,000 psi, far in excess of the expected yield point. It is apparent that the slab cracked slightly under the excessive dead load, thus affording access from the underside of the slab for corrosive air. The corrosion was highly localized; the wire 1 in. away from the break was perfectly clean.

Even the impact and added load of the slab, which having split down the middle, fell so that it rested diagonally on the second floor near the center of the span, did not cause failure of the tile-and-joist floor.

There is little doubt of the fact that such a failure from corrosion would not happen today. Even the best concrete of 1906 was probably not as good as the poorest we would get today. Certainly we would not stand for a mixture with as much water and therefore as much porosity as the concrete which, in those days, was considered a good mix.

The Campanile—Venice, Italy

After more than half a century the sensation has abated somewhat, but the world-famous collapse of The Campanile [141] on St. Mark's Square, Venice, July 14, 1902, is still a matter of more than historical interest.

The famous old tower was begun in 888, in the early days of Venetian glory. It was not completed, however, until 1517. It was founded on stone, resting on hemlock piles, which were capped with several courses of oak timber about 20 ft below the grade of St. Mark's Square and several feet below low tide. The subsoil is clay, with streaks of sand and gravel to great depth. An examination of the foundation in 1885 proved it to be sound. Several contributing causes to the collapse were suggested—dredging of the canals nearby, minor earthquakes in northern Italy, and, most recently, an earthquake at Saloniki.

An official report issued later stated that the lower part of the tower was built of grouted filling between inner and outer masonry walls, that numerous repairs had been made in the seven or more centuries it had stood, and that much of this repair work was inadequate. As a result, the grout had gradually come to carry most of the load, and repair work, undertaken in the outer wall a few days before the failure, produced cracks which precipitated the collapse.

As a result of earthquake shocks it had experienced through the years, the Campanile had tilted so that it was 3 ft out of plumb at the top.

In the reconstruction work that followed, the new Campanile was built as a duplicate of the original building. However, 4,000 new piles were driven to support the foundation.

9.4 Overload

The failure of buildings through overload is not necessarily confined to old buildings, although these often fail because, throughout the years, they have carried heavy loads—probably heavier than they were designed for—and the owner or occupant has come to believe that his building will carry whatever load he wants to put on the floors.

Building codes nowadays require the use of certain design loads for certain types of buildings, but this does not necessarily mean that the owner or user of the building will limit the floor load to this weight. The writer recalls an instance where, despite the fact that the building code requirement for heavy warehouse use was 150 psf, he made the design for 300 psf, knowing that it might be used for fairly heavy storage. Some time later on an inspection trip, he checked this area and found that it was carrying sheet lead to an average load of 450 psf. Fortunately, no failure occurred. In former days, before building departments existed, the floors were built more or less by the rule of thumb to carry the intended loads. Frequently the use of the building was changed and the applied loads were greatly increased, as in the Park Place building.

The overload that causes a failure can be a direct overload of a floor or may be a lateral overload on a wall.

Morris Fruit Co.—Minneapolis, Minn.

On October 27, 1937, two men were killed and thirteen were injured when a portion of the third floor of an old building collapsed. It was used as a warehouse and office building [142] by the Morris Fruit Co. in Minneapolis, Minn.

The failure was almost certainly caused by overload of the floor. Although the building was 41 years old, neither weakness or deterioration is indicated as a contributing cause of the failure. The *Engineering News-Record* described it as follows:

The building, in the market district of the city, is a three-story solid brick wall structure about 40 × 100 ft in plan. It was built 41 years ago, in 1896. The interior floors were supported by a row of columns down the center of the building, spaced 20 ft apart, and carrying a longitudinal girder, dividing the floor into bays 20 × 20 ft. Joists spanning across the main girder and spaced $3\frac{1}{2}$ ft apart were built up wooden beams, 7 in. by 12 in. The crosswise joists were set in the side brick wall about 6 in., giving a bearing area of about 40 sq in. A 2-in. rough floor was laid on these joists, and over that a finished floor.

The righthand side of the first and second bays, viewing the building from the front, were unloaded. On the lefthand side, the first two bays carried a load of about 240 psf, consisting of 52-lb cases of ginger ale, stacked six and seven high. The safe load for this floor, as figured by the Minneapolis Building Inspector, is 90–100 psf.

In the failure, the front bay on the loaded side gave way. The load shifted toward the front wall, and it appears that the joists broke at about the quarter-point from the wall. The second loaded bay did not fail and held its load.

The panel which fell broke through the second floor and damaged the first floor but did not go through to the basement. The first and second floors are used as offices by the fruit company.

None of the columns and girders failed and, except for one panel which dropped, the floors of the building are intact.

Bienville Street buildings—New Orleans, La.

In the case of the failure of an old building, it is sometimes difficult to say whether it resulted from the age and deterioration of the building or from the loading of the floors beyond the capacity of the floor joists when they were new. Two old buildings[143]—each with 30-ft frontage and 80 ft deep, on Bienville Street, New Orleans, La., collapsed without any warning, July 23, 1928. The buildings were probably almost 100 years old, of brick construction with lime mortar, having 16-in. walls in the ground floor and 12-in. walls above. Cast-iron street-front columns carried door and window lintels. The roof and upper floors were of wood construction; the first floor of concrete on the ground was placed at a later date. Originally the first floor had been occupied by stores and the second and third floors by residences. Before the collapse, one building was vacant throughout, and the other was used for the storage of leather goods, and raw leather.

Apparently, either the overload on the upper floors, or the lateral thrust caused by the leather against the party wall between the two buildings, perhaps both, kicked over the party wall and the wall opposite, wrecking both buildings.

Wansco Paper Products Co.—New York, N.Y.

Parts of New York, N.Y., and almost every big city in America are full of old buildings that could fit this story.

On December 27, 1944, an old five-story building,[144] of the type built in the post-Civil-War era, collapsed. The failure took place in the evening and no casualties occurred. The building had a five-story ornamental cast-iron front, made up of cast-iron columns and panels. The side walls were brick bearing walls, interior columns were of cast iron, and the rest of the framing was of wood.

Cases of paper goods—piled ceiling high on the fourth floor—provided the overload to precipitate the failure. The entire interior of the building collapsed, and the cast-iron front of the three lower stories failed. The two top stories of the cast-iron front stayed in place, apparently bridged between the structure on either side.

Center Street warehouse—Newark, N.J.

On March 3, 1921, the entire rear wall of a five-story warehouse [145] in Newark, N.J., buckled outward and fell. Aside from property damage, there was no report of any casualties.

The owner leased his building to a tenant who overloaded the floors with waste paper stock to such an extent that the building failed. The owner had warned the tenant that he was overloading the structure, but the lease called for the tenant to repair any and all damage incurred so the landlord neglected to make any report of the overload to the building department.

Dye plant—Philadelphia, Pa.

Building departments should require all floor areas used for manufacturing or storage to have the permissible live load posted on the walls or columns of the area, and management should be responsible for periodic checkups to see that these limits are not exceeded.

On August 1, 1912, the second floor of a dyeing plant [146] in Philadelphia, Pa., collapsed under a load of 30,000 lb of yarn stored there. In itself, this is a comparatively light load but, concentrated on a small area, it could easily have precipitated the failure.

The structure was a three-story building with wood floors and brick walls. The failure disturbed the supports of a 10,000-gal water tank, and this tank fell, along with an 80-ft smokestack, completely wrecking the building, killing two men, and seriously injuring two others.

Residence—Erie, Pa.

In all probability, some readers may wonder why this type of failure does not occur more often. On February 28, 1911, the floor of the kitchen of a residence [147] at Erie, Pa., failed during a funeral. The room was crowded with fifty mourners at the time, and eight of them were seriously injured when the kitchen floor collapsed into the basement.

Residence—Aurora, Ill.

Group gatherings in houses for weddings or for other reasons may prove as disastrous as those for the funeral which has just been described. It is not the reason for the gathering, but the extent and nature of the floor load, that may cause trouble.

On January 23, 1914, a stairway and floor in a residence [148] in

Aurora, Ill., collapsed during a wedding celebration dance. Twenty-two of the one hundred guests were injured—the bride and groom among the most seriously.

Huey and Philps Hardware Co.—Dallas, Tex.

When a section of the end wall of a large brick warehouse at the Huey and Philps Hardware Co.[149] in Dallas, Tex., collapsed, July 10, 1914, the failure was like the fall of a row of dominoes, each knocking over the next one.

A series of wooden racks, located as shown in Fig. 9.6, were used to carry rods on the second floor of the building. The tracks of the Texas and Pacific Railway ran along the side of the building. It was thought that vibration caused by a heavy freight train and prob-

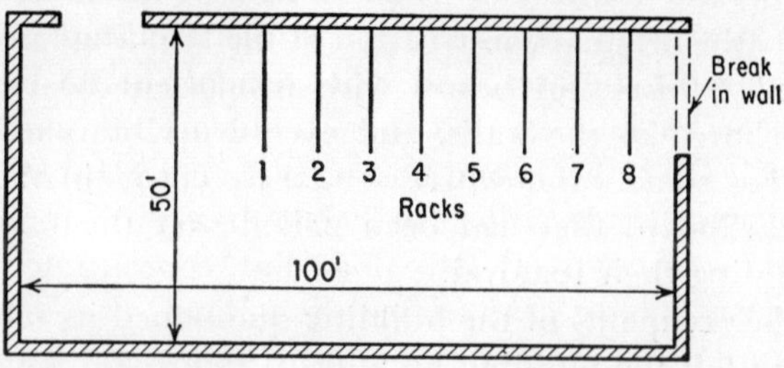

FIG. 9.6. Plan of third story—Huey and Philips Hardware Co.

ably aided by some eccentric loading of Rack No. 1 tipped it over against Rack No. 2, which in turn tipped Rack No. 3, and so on until the cumulative force, probably aided by rolling steel, hit the end wall and broke out the brickwork. Fortunately no one was at work in the building at the time so no casualties occurred.

Apple warehouse—Chambersburg, Pa.

An engineer might well wonder how to go about computing the lateral pressure developed by a pile of apples.

A cold-storage warehouse [150] in Chambersburg, Pa., constructed with 8-in. concrete and 4-in. brick walls, was lined with cork insulation. The free-standing walls were tied to the reinforced-concrete floors with ¾-in. anchors, 10 ft on centers.

The warehouse was filled with apples from the recent harvest, and in November 1942, when the anchors straightened out and a section of wall 50 × 110 ft fell under the lateral pressure of the apples, 15,000 bushels of apples rolled onto the streets and railroad tracks.

No casualties occurred, and the floors, which were carried independently by the concrete columns, suffered no damage.

Park Place building—New York, N.Y.

One of the worst failures on record caused by overload on an inadequate construction, occurred August 23, 1891, when the collapse of a building [151] on Park Place in New York, N.Y., killed 64 people.

The 20-year-old building was in questionable condition for some time because of constant overload and had been inspected by the building department not long before the failure. It had been designed for relatively light loads, but, several years before the failure, the fourth and fifth floors had been occupied by a printing company that installed fifteen heavy presses, a dead weight of 50 tons that had the added danger of oscillation in the weak direction of the framing, together with a considerable tonnage of paper to feed the presses.

Apparently the original construction of the building left a great deal to be desired. Joists and beams with insufficient bearing were not properly anchored to the walls, and excessive vibration had cracked the walls. For some unknown reason, these cracks in the walls were hidden by cardboard that had been nailed over them and carried a notice forbidding their removal.

Some of the occupants of the building questioned its safety, and several workmen left the building because "it shook so." A building inspector "found nothing to excite his suspicion" and approved it for its use as a printing establishment.

Apparently the fall of a 600-lb bale of paper from the top of a pile was the final cause that triggered the failure.

Whitney Noyes Seed Co.—Buffalo, N.Y.

Building owners often have the peculiar notion that the safe load for a floor is limited only by the cubic capacity in which the material may be piled.

On January 25, 1910, the three upper floors of the Whitney Noyes Seed Co.[152] in Buffalo, N.Y., failed—fortunately with no injuries to anyone. The building was a well-built, four-story structure of timber floor construction and brick bearing walls. The upper floor had been designed for a live load of 130 psf, and the ground floor was designed for 225 psf. The permit for construction had been issued in 1896.

An area one bay wide, about 15 × 50 ft in area, of the three upper floors collapsed, breaking out about half of the front wall. The load on the three upper floors in the bay that had not failed was found to be from 200 to 400 psf. Under the circumstances, there was no reason to search for the cause of the failure.

Playhouse Theater—Wilmington, Del.

An occasional check on the condition of roof drains should be part of the regular maintenance checkup system, particularly on any public building. Falling leaves, building up to form the stoppage of a roof drain, can hold back water to build up a load in excess of the capacity of the roof.

The roof over the stage of the Playhouse Theater [153] at Wilmington, Del., collapsed during a performance on the evening of December 11, 1925, because of a heavy accumulation of roof water resulting from a stopped-up roof drain. Nobody on the stage at the time was seriously injured.

Washington Building—Buffalo, N.Y.

Although this may appear to be a concrete failure, it does not seem fair to list any shortcoming of the material as the cause, when it has done so much more than should be expected of it.

On February 28, 1924, a section of the roof of the ten-story reinforced-concrete Washington Building [154] in Buffalo, N.Y., failed—breaking one panel on the top floor and coming to rest on the ninth floor without damaging any part of it. The floors were designed for 120-psf live load, and the roof was designed for 40-psf snow load. The concrete had been poured on January 15, and, since the weather had been cold, salamanders had been kept fired until the windows were installed, and a steam heating system was in operation.

On the afternoon before the failure, the coping stones and barrels of tar had been piled in an area of about two or three bays on the roof, loading the roof area in question to between 150 and 200 psf. The following morning, the lather, who was to install a hung ceiling over the tenth floor asked one of the foremen to strip the shores to permit him to start. The foreman, not knowing of the heavy concentration of load on the roof, proceeded to remove the shores, and the collapse occurred, injuring one man. When a slab designed to carry 40 psf fails under a load of 150 to 200 psf, it can hardly be the fault of the slab.

Royal Canadian Hussars Armory—Montreal, Que.

Normally, a new roof designed to carry a snow load of 40 psf will stand in almost any place in United States or Canada, even though the factor of safety may be slightly overworked, but occasionally some particularly heavy snowfall may strain even this limit.

The armory being constructed for the Royal Canadian Hussars regiment at Montreal, Que., was almost completed, when an unprecedented snowfall built up about 6 ft of snow on the roof, probably about 120 psf on a roof designed for 40 psf, and caused the collapse of about 5 per cent of the area in the early morning of February 19, 1936. No casualties occurred.

The main support of the roof was formed of three hinged steel arches, none of which failed. On this were light steel purlins supporting a timber deck. Under the circumstances, it would seem that the excessive snow load that caused the failure could be classified as an "Act of God," thus absolving anyone from failing to design or build properly.

10

Wind, Fire, and Explosion

Although it is customary to term windstorms, fires, and similar disasters as "Acts of God," it is often true that some simple precautions could have averted building damage.

10.1 Windstorms

The results of windstorms during construction operations have already been noted in Chap. 3 and 4. None of the structures destroyed in these cases was the result of an extraordinary windstorm, but rather of windstorms that are experienced throughout the country every year. They are a hazard of the construction industry, and insurance adjusters feel the contractor should provide for them in the ordinary course of his operations.

Tornadoes and hurricanes, however, are a different matter. Tornadoes are freakish windstorms that do a great amount of destruction in a relatively narrow path in a very short space of time. So far, no type of man-made construction has been able to withstand the pressures of a tornado. The eye of a tornado usually has a width of less than one-half mile.

The hurricane, on the other hand, is a cyclonic storm with a diameter of 100–300 miles; although hurricanes are much more harmful than ordinary gales, the results of the Florida hurricane of 1926 have proven that properly designed and built modern buildings will stand up relatively well.

In general, building codes are not much concerned with the strength of buildings during construction; their sole concern is for the safety of the building after it is occupied. It is generally understood that it would work a needless hardship on the owner to require him to build a structure to resist full hurricane winds. The statistics of the Factory Insurance Association would seem to bear out this decision in that claims for windstorm losses show that 76 per cent of all claims are classified under the heading of "high winds," 19 per cent as "hurricanes," and 1 per cent as "tornadoes," with the balance being "hail" or "exposure." Even with hurricanes, however, the damage can be cut down tremendously by reasonable precaution.

After Hurricane Donna hit Miami in 1960, a five-man team of engineers investigated the damage done to buildings in the area—damage that hit residential buildings primarily.[155] As a result of their inspections they estimated that an added average cost of $25 per home would have provided the margin of safety necessary to withstand such a hurricane. The three primary weak points indicated by their investigations were: (1) the connection of the anchor straps from roof rafters to plate by means of two 8d nails instead of three 16d nails as required by the County Building Code; (2) the use of tarpaper joint over the cores of a concrete masonry wall to keep the concrete of a poured concrete lintel from filling the blocks; and (3) the anchorage of 2-in. tongue-and-groove sheathing to supporting trusses by means of a single 30d nail. Every one of these recommendations was a simple question of proper anchorage.

In selecting the few failures that have been written up, an attempt has been made to emphasize the type of windstorm failures that might be avoided. Two office buildings in downtown Miami failed in the hurricane of 1926, whereas a number of others, equally tall and otherwise equally vulnerable, suffered only local damage. Why?

Hurricane—Miami, Fla.

The safe design of tall buildings to withstand wind pressure was entirely a matter of theory until the hurricane of September 18, 1926, in Miami, Fla.,[156] demonstrated to engineers the sufficiency of their designs. We know, of course, that those buildings in New York and Chicago and other large cities which had been wind braced, had stood up for years under the buffeting of the tempests, but we had no idea how much excess strength we had built into them. Then, probably, the worst winds that had ever hit the center of a modern American city gave us proof of the soundness of this theory. Under winds of 125-mph velocity, blowing first from one direction, then reversing and blowing from the opposite direction, only 2 out of 25 or 30 modern, tall, framed structures of steel or reinforced-concrete yielded. To be sure, many of them suffered much damage. Window and wall panels were blown in or out, but after it was all over, only two of them had suffered any damage to their structural framework—the Meyer-Kiser Building and the Realty Board Building. Fig. 10.1 indicates their location in Miami.

The Meyer-Kiser Building, Fig. 10.2, suffered the more severe damage of the two, being distorted about 2 ft toward the west at the south wall, and 8 in. toward the east at the north wall, as indicated in Fig. 10.3. It was a seventeen-story building, with floors of concrete joists and tile

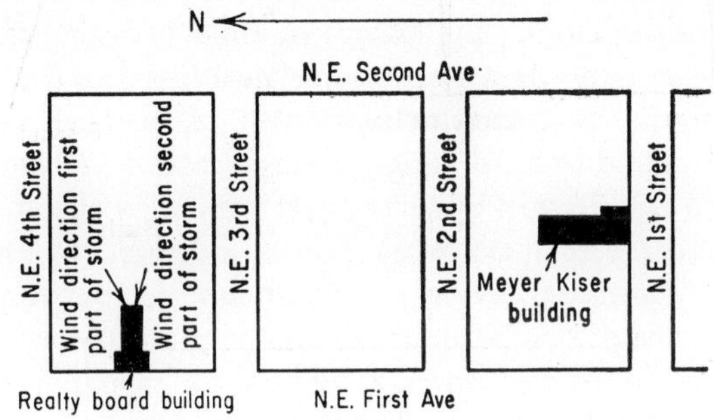

FIG. 10.1. Location of buildings during 1926 Miami hurricane

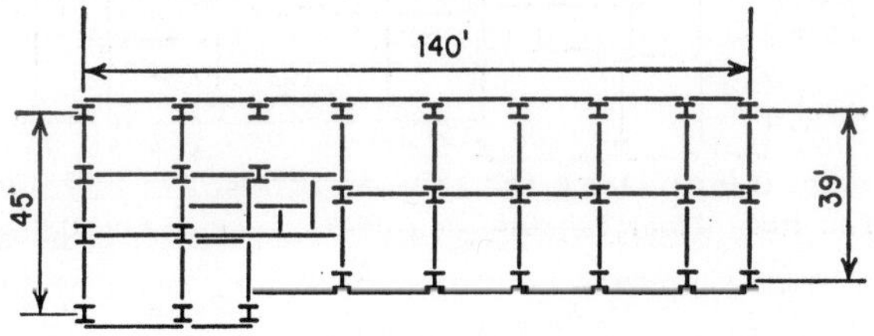

FIG. 10.2. Floor framing plan of Meyer-Kiser Building

fillers as shown in Fig. 10.2. These floors performed as they should have performed, transferring their stresses as horizontal plates to the vertical steel wind bents of the building. The columns through the lower four or five floors remained practically plumb, but from the fifth to the thirteenth floor they were badly distorted. Above the thirteenth floor, they again assumed a plumb line. Apparently the front masonry wall had yielded first through the distorted stories, and the steel without the assistance of the masonry was unable to stand the strain. It was decided that it was impractical to repair the entire building, so the upper ten stories were taken down, leaving a seven-story building instead. As the upper stories were taken down, parts of the frame attempted to assume their original vertical position, springing back as

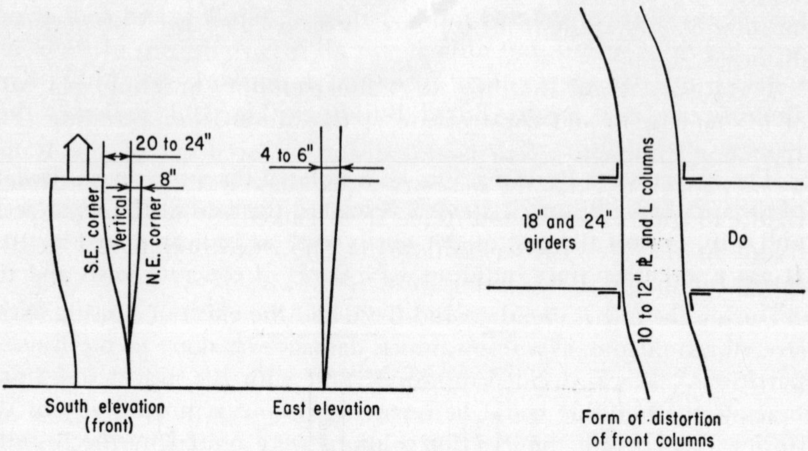

Fig. 10.3. Distortion of Meyer-Kiser Building

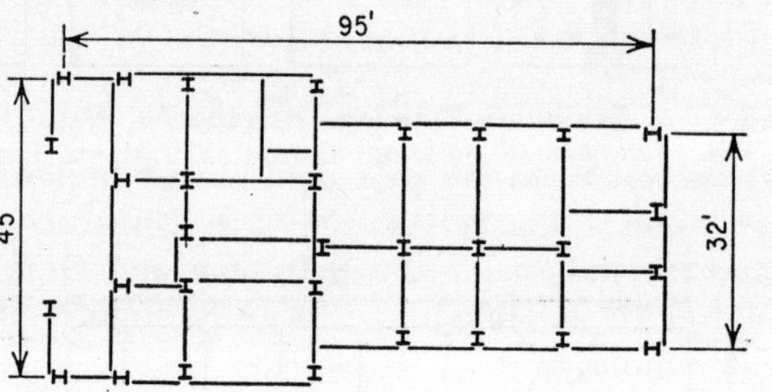

Fig. 10.4. Floor framing plan of Realty Board Building

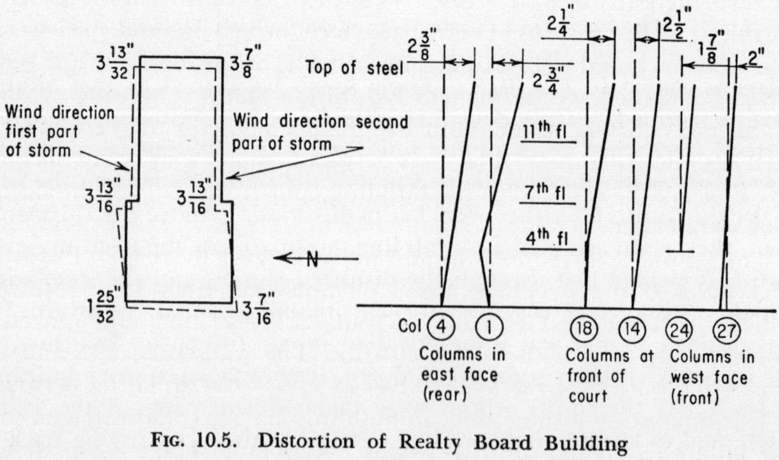

Fig. 10.5. Distortion of Realty Board Building

much as 3½ in. The front columns, however, had been permanently distorted.

The Realty Board Building was a fourteen-story building, generally of the same plan and construction as the Meyer-Kiser Building (Fig. 10.4) but, although severe lateral distortion occurred in the frame, severe enough to crack the masonry walls, these walls were apparently rigid enough to prevent the frame from exceeding its yield point, as it did in the Meyer-Kiser Building. The distortion of the Realty Board Building is shown in Fig. 10.5.

During the hurricane, the wind force had the effect of causing excessive vibration, and, as a result, much damage was done to the interior partitions. It was decided, however, that with the aid of some new bracing, the building could be repaired without removing any of the stories. Several of the exterior columns were braced by the installation of reinforced-concrete spandrels from the window head of one floor to the window sill of the floor next above, up as far as the sills of the eighth floor. All of this work was done to provide structurally to resist a wind force of 20 psf as required by the Miami Building Code.

Baltimore and Ohio Exhibition Building—Halethorpe, Md.

It is hardly necessary to say that light-weight buildings are not resistant to high winds. The very nature of the construction—high walls, light roof, and the like, made the Baltimore and Ohio Railroad Exhibition Building [157] at Halethorpe, near Baltimore, Md., a particularly vulnerable building.

It was built in 1927 for the B. & O. "Fair of the Iron Horse," and it may have been intended as a relatively temporary building, even though built of nonflammable materials. It was a building 60 × 600 ft, 18 ft high at the walls, which were of brick construction, mostly 8 in. thick. The roof was of corrugated steel on purlins and steel trusses that were anchored to brick piers in the walls by means of anchor bolts. It was being used as a permanent Exhibition Hall for housing exhibits of railroad equipment, including original locomotives and cars, models, sketches, and pictures. The only large openings were large locomotive doors at the east end of the building, and at the time of the failure they were closed.

The failure occurred in a thunderstorm, August 13, 1935. The wind that wrecked the structure was a 60-mph gale from the southwest, which lifted the west end of the roof off by suction and moved it south, lifting part of the walls with it. The walls then fell, and the trusses fell on top of the locomotives or cars. No casualties occurred.

It is interesting to note that, in buildings having a light-weight roof, the failure in a windstorm is usually precipitated because the roof of the structure is lifted off by suction not by the walls blowing over.

Daytona Motor Co.—Daytona Beach, Fla.

That a building can comply with all the codes, can be built under the strictest supervision, and still fail, was shown in the failure of the parking garage [158] of the Daytona Motor Co., Daytona Beach, Fla., during a storm in the fall of 1936.

The structure that collapsed was a combination parking shed without side walls, 105 ft deep by 102 ft wide, in front of a partially enclosed repair shop 111 ft deep by 102 ft wide, which was added at a later date, as shown in Fig. 10.6. The extension consisted of a series

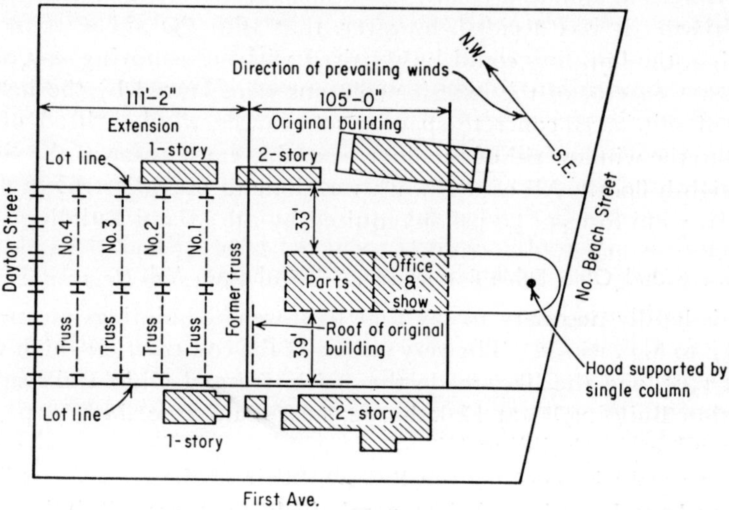

FIG. 10.6. Service station, car parking shelter, and adjacent buildings

of trusses carried on wall columns and center columns across the building. Around the two sides and back of the building a concrete masonry wall formed a base 4 ft 9 in. high, and a metal apron 6 ft high extended down from the roof, leaving the intermediate space open. The roof was of 2 × 6 tongue-and-groove lumber, spiked to wood nailers on steel beams. All members throughout were designed in accordance with good practice, were well-braced, and were capable theoretically of carrying a total load of 40 psf.

The first theory put forth for the failure was the accumulation of rain on the roof, but this was impossible because the roof had no parapets, only a 2-in. gravel stop around the edge. The water could only accumulate to that depth.

The final conclusion of the engineers as to the cause of the failure was that the wind blowing through the open face of the building collided with counter currents from the side and from the rear, met in a gusty condition and deflected upward to lift the center column under Truss 4 off its base, then drop it down again, thus precipitating the failure.

10.2 Fires

Although many important lessons may be learned from the so-called "conflagrations" of history, such as the great Chicago fire of 1871, and the San Francisco fire following the earthquake of 1906, any extended discussion of these fires would be useless. In the first place, most of the lessons to be learned have already been recognized. Then too, modern building codes and insurance regulations forbid construction that would contribute to such conflagrations. Moreover, we have had no such fires in recent years, so that any discussion of them would be largely a matter of historical rather than technical interest. It might be interesting, however, to list these great fires of the past 100 years, simply to remind us of what the application of the lessons learned has done for the new construction of today. I have selected the following from a publication of the Richards-Wilcox Company, manufacturers of automatic fire doors.

1. July 4, 1866, Portland, Me., one-half of the city, $10 million loss
2. September and October, 1871, forest fires, Michigan, Minnesota, Wisconsin, 759 sq miles, 17,430 buildings, $11 million loss
3. October 9, 1871, Chicago, Ill., 2,000 acres, 776 buildings, 250 dead, $165 million loss
4. November 9, 1872, Boston, Mass., 65 acres, $75 million loss
5. June 6, 1889, Seattle, Wash., business section, $6 million loss
6. October 28, 1892, Milwaukee, Wis., 600 buildings, $9 million loss
7. May 3, 1901, Jacksonville, Fla., 148 blocks, $10 million loss
8. February 9, 1902, Paterson, N.J., 456 buildings, $5 million loss
9. February 7, 1904, Baltimore, Md., 140 acres, 2,500 buildings, $50 million loss
10. April 18, 1906, San Francisco, Calif., 2,593 acres (4.05 square miles), $350 million loss
11. April 12, 1908, Chelsea, Mass., 492 acres, $6 million loss
12. April 30, 1911, Bangor, Me., $3 million loss
13. January 9, 1912, New York, N.Y., Equitable Building, $3 million loss
14. February 21, 1912, Houston, Tex., $4 million loss
15. June 25–26, 1914, Salem, Mass., 1,700 buildings, 253 acres, $14 million loss

There is no such thing as a fireproof building. A building may be built of nonflammable materials, but as long as there are flammable materials stored within the building, or flammable furnishings or trim, the building can only be called fire-resistant. In the minds of the public, any building that cannot be consumed by the fire is fireproof. In the case of the La Salle Hotel in Chicago, Ill., and the Winecoff Hotel in Atlanta, Ga., the buildings were intact after the fire, although several hundred lives were lost in the two fires. Of the thousands of fires that occur annually throughout the United States, only a few are written up here, those that may teach the best lessons to modern engineers. Such classic examples as the Iroquois Hotel fire in Chicago, Ill., in 1903, in which 564 people died, are omitted, because the faults of design which brought about the loss of life have been largely legislated out of existence since then.

Although they do not involve the destruction of a finished structure, several construction fires involving formwork and scaffolding are included here, not so much because of what we may learn from them as out of general interest.

Reinforced concrete does not burn, but this does not mean that it is fireproof. Subjected to excessive heat, the concrete will disintegrate and the building will be ruined. This excessive heat may be provided by the contents of the building if there are flaws in the fire protection system or in any of its many details. In the earlier days of reinforced-concrete construction a series of three such fires within a relatively short time 45 years ago provided the factual information that has enabled us to reduce greatly the damage from such fires since then.

Edison Plant—West Orange, N.J.

The first of these fires occurred in the plant of Thomas A. Edison, Inc.,[159] West Orange, N.J., December 9, 1914, when a group of structures of various types of materials burned. The fire started in the film-inspecting department, a one-story wood-frame building, and, because of the highly flammable contents, it spread rapidly to adjacent buildings.

The fire was partially checked by the concrete wall and structure of several reinforced buildings, but this was only temporary. The wood sash permitted the entry of the flames, and the fire spread rapidly through the flammable contents of the buildings, which housed such operations as cabinet making, phonograph assembly, and varnishing.

Without going into a complete survey, structure by structure, it may be stated that within the area of the conflagration there were about sixteen buildings, seven of which were of reinforced-concrete, three to five stories high. All of these reinforced-concrete buildings were of beam-and-girder-type construction, and all but one of them had square or rectangular columns. The three upper floors of one of these buildings collapsed, and the roof of two others collapsed. Throughout the rest of the buildings, the concrete spalled off under the heat—particularly on the corners of the columns and beams. The presence of fused metal in various places indicated that the fire had reached an intensity of 1,000°F in all of the buildings, and in excess of 2,500°F in several places. Under the circumstances, the concrete stood up remarkably well. Of 1,360 column sections involved in four of the buildings, only 103 of them failed. After the fire 947 of the sections were completely intact.

An interesting contributing cause of damage was that expansion bolts and other steel connections for the suspension of piping and wiring were embedded in contact with the steel reinforcing bars, forming a direct channel for the high temperature into the beams, thus bringing about spalling of the beam soffits.

The report of a committee of prominent architects and engineers submitted on the Edison fire, criticized the following deviations from well-established practice of 1914.

1. Large, undivided floor areas—lack of fire walls and stops
2. Unprotected exterior openings—such as wooden sash with plain glass
3. Absence of proper fire walls, partitions and fire doors
4. Failure to properly enclose vertical openings, stairway and escalator shafts
5. Absence of adequate sprinkler equipment
6. Absence of adequate water supply and fire-fighting apparatus
7. Failure either to isolate nonfireproof buildings or to protect against them
8. Failure to isolate hazardous processes of manufacture, either by placing such processes in separate buildings or surrounding them by proper fire walls

Repairs were made where feasible after the fire, and load tests following the repairs indicated that the repaired floors were capable of carrying their original design load.

Mullen and Buckley warehouse—Far Rockaway, N.Y.

The second of these fires occurred in the seven-story-and-basement Mullen and Buckley warehouse[160] at Far Rockaway, N.Y., November 10, 1916. The point of greatest interest in connection with the fire was the serious destruction of concrete under fire conditions of comparatively minor intensity.

The fire began in a frame building next door and found access to the reinforced-concrete warehouse through a 5-in. hole, used for the passage of a carrier system. Immediately above this point of access was a 4-in. hole in the floor left by the removal of a heating pipe. The building was mostly filled with flammable furniture, which fed flames so that they spread rapidly. A poorly fitted fire door with ½-in. openings around it permitted the fire to spread to other parts of the building.

The building was a well-built structure in accordance with the best known standards of the times, and the evidence was that in most places, the effect of the fire was of short duration, but the damage amounted to about $125,000.

The failure consisted largely of localized cracks in walls and parapet, the collapse of three panels of the fourth floor, the failure of two column sections, the deflection of several beams, and bad spalling of the concrete beams and columns. The building was completely repaired and restored to use late the following summer.

The report, by Ira H. Woolson, consulting engineer for the National Board of Fire Underwriters, became a classic in the determination of fire codes for reinforced-concrete construction. The major point developed was that a sprinkler system would have saved the warehouse and probably the frame building. The rapid surface expansion caused by the heat was a major cause of spalling, and this condition was made worse by the sudden cooling when the fire streams hit the hot concrete. The exposure of the steel reinforcement, as a result of the spalling, permitted a softening of the steel and deflection of some of the beams as a result. The concrete had been made using a quartz gravel aggregate, which disintegrated badly from the heat.

Quaker Oats Co. warehouse—Peterboro, Ont.

The third of the reinforced-concrete building fires in this sequence occurred at Peterboro, Ont., December 11, 1916—a fire in which 26 men lost their lives.[161] Although the fire destroyed nine buildings, causing a loss of $2 million, most of the buildings destroyed were of a semi-mill-type construction, in which the timbers used in the building contributed to the fire. One of the buildings—the warehouse to which this article pertains—was of reinforced-concrete flat slab construction, a system then, and even now, accepted as probably the most modern fire-resistant system possible. It had round columns, very few beams, and relatively few of the square corners that caused such serious spalling in the fires described previously.

Disregarding the fires that destroyed the mill construction buildings, we may say that the fire started at 10:00 A.M. from an explosion in the drying building, a building with brick walls, steel beams and columns, and concrete floors. All of the buildings were interconnected, either directly, or by bridges and tunnels. All buildings except the reinforced-concrete warehouse, the reinforced-concrete grain elevator, and portions of the drying building where the fire began, were protected by a dry-pipe, automatic sprinkler system. The warehouse had been erected as a four-story warehouse in 1910, and the upper two stories were added in 1916, at which time the columns of the lower floor were strengthened to carry the additional load. The original four stories, of flat-slab construction, had octagonal columns and no drop panels. The two additional stories had round columns and drop panels. The entire building was protected with wire glass windows. The concrete throughout was of excellent quality, using a local gravel as aggregate.

The explosion blew out several walls of the building, and these falling walls broke the steam lines operating some of the fire pumps, thus preventing their full use. The fire spread rapidly, fed by the cereal contents of many of the buildings, and reaching the warehouse within 45 minutes. Here large piles of wood used for the manufacture of boxes, furnished the fuel. Without adequate fire-fighting facilities, the fire burned itself out in the warehouse when about half of the interior construction collapsed. The remainder was so badly damaged by the spalling of columns, the sagging of floors, and cracking of walls that, to all intents and purposes, the building was a complete loss.

The intensity of the heat developed may be judged by the fact that the brick used in the walls was fused in many places. It is estimated that the fire had almost blast-furnace intensity, reaching a temperature of more than 2,200°F.

Chamber of Commerce Building—Cincinnati, Ohio

To the layman nonflammable may mean fireproof, but to the initiated this is not true. Steel may be nonflammable, but unless it is protected against the possibility of being heated to the point of softening up, it is not fireproof in the sense that the word is sometimes used in the building industry. As a result of what might have been a relatively minor fire hazard had the structure been properly protected, the Chamber of Commerce Building,[162] an architectural landmark in Cincinnati, Ohio, was completely destroyed, June 10, 1910—not so much by fire as by the collapse that resulted from the fire.

The building was designed by the famous Boston architect, H. H. Richardson, and was completed in 1889. It covered an area of 100 × 150 ft and was 200 ft high. The exterior walls were of granite, ranging from 2½ ft thick at the top to 6½ ft thick at the base. The ground floor was used for banking and over it the Exchange Hall had a clear and unobstructed area of 66 × 136 ft, 50 ft high. In order to accomplish this, the upper floors were suspended from steel roof trusses in the attic space. The floor beams throughout, originally designed of wrought iron, were changed to steel, thus making it the first steel building in the area.

Over the Exchange floor three floors were used for office purposes, and the top floor was occupied by the Business Men's Club for dining rooms and club kitchen. In the attic space above the kitchen (which had never been intended for use), was an auxiliary kitchen for the club's help and storage space for ice chests and general storage.

It is thought that the flames from a grease fire in the kitchen escaped through a leaky flue to the junk stored in the attic space, which almost immediately became a raging furnace. The unprotected steel trusses softened from the heat, slowly yielded and, within 15 minutes, even before the firemen could get a stream on the fire, the entire center of the structure had dropped to the basement, killing seven people. The floors throughout were of tile arches, supported on protected steel beams, and the building was supposed to be a fireproof building.

LaSalle Hotel—Chicago, Ill.

Among fires of interest to architects and others outside the insurance field is the LaSalle Hotel [163] fire in Chicago, Ill., in the early morning of June 5, 1946.

The LaSalle was a 1,000-room, 22-floor hotel, of fire-resistant construction, with a steel skeleton encased in concrete, concrete floor slabs, and with brick and stone curtain walls. It had been completed in 1909 and, at the time it was built, was the last word in safe construction. At that date stair enclosures were not required. The building had two open stairs from the lobby to the twentieth floor, and three enclosed stairs—two from the twentieth to the second floor and one from the twentieth floor to the lobby. Two outside stairs furnished the only safe exit from the upper floors because of the rapid spread of the fire.

The fire started on the ground floor and, fed by the flammable contents of the floor, spread rapidly through the lobby to the elevator shafts and stairwells, sending clouds of smoke and gases to the upper floors, so that none of the inside stairs were available for escape. The fire caused extensive damage to the five lower floors, caused 61 deaths, and more than 200 other casualties.

The above short paragraph tells, briefly, what happened, but it requires more detail to develop fully the lessons to be learned from this holocaust. The building frame, the structural floors, and the brick walls were still in sound condition after the fire had burned out—the building itself was fire resistant. However, the trim, particularly in the first floor, was highly flammable. The room doors on the floors above were wood panel doors with wood transoms. The lobby and mezzanine had beautiful walnut-veneer paneling mounted on wood furring with concealed spaces behind the paneling. Tests of the walnut paneling after the fire showed that the plywood paneling would almost immediately delaminate when heat was applied, and that the delaminated veneer would spread the flames at a rate five times that of the surface of an ordinary solid red-oak panel.

The most vulnerable part of the entire building, however, was the cocktail lounge where the fire originated. It was a very attractive wood-studded room within a room, all of the trim being highly flammable. The center of the room had a circular bar, and wood tables with built-in upholstered wood benches were arranged around the walls. Ventilation for this room was furnished by a 3-ft-diameter hole through the masonry wall of the elevator shaft, using the shaft as a flue. Another small hole from the elevator shaft opened into the space behind and under the wood benches; it is possible that a cigarette butt or match thrown through the open cage of the elevator lodged in this area and started the fire. Above the elevator doors at the first floor level, the wood paneling had no masonry behind it.

The inquest that followed the disaster developed the story of red tape "proper channels" rules, which, although they had nothing to do with the physical properties of the building, are of interest. Inquiry into the reason for the delay in turning in the alarm brought out that employees of the hotel were not permitted to call the fire department without first reporting to their superiors, who in turn reported the fire to either the manager, the house detective, the chief engineer, or the chief electrician. These men were authorized to call the telephone operator, who transmitted the alarm to the fire department if she recognized the voice of the caller as one in authority. In this instance, the hotel detective authorized the telephone operator to call the fire department about fifteen minutes after the smoke had first been noted by a patron of the cocktail lounge. The operator, after warning the occupants of the upper floors, died of asphyxiation, trapped behind the barred windows of her second floor switchboard room.

Winecoff Hotel—Atlanta, Ga.

The burning of the 31-year-old, 15-story Winecoff Hotel [164] in Atlanta, Ga., in the early morning hours of December 7, 1946, was the second such disastrous hotel fire in the United States within 6 months.

The fire apparently started in a fifth-story corridor and had gained considerable headway before discovery. The open stairwells, without fire door shutoffs, permitted the flames to spread rapidly to the upper floors, and the small amount of flammable material in the carpets, painted surfaces, wooden room doors, and the like, fed the fire enough to cut off escape from the upper floors completely. Within ten to fifteen minutes of the time the fire was discovered, guests were jumping from the upper windows. Loss of life on the upper floors was appalling. The majority of those who attempted to escape through the hallways died from heated air. Others died in their rooms or jumping from windows. Of the 304 guests who were registered at the time of the fire, 65 were injured, 119 died, and 120 escaped injury—a higher proportion of casualties than any other hotel fire.

Although the LaSalle and the Winecoff hotel fires were similar in that both were rated as fireproof buildings and neither suffered any structural damage, the direct causes of so many deaths were different. The primary reason for this disaster was the existence of only one stairway. Because of the arrangement of this stairway, the fire did not spread so rapidly from floor to floor. The two elevator shafts were enclosed in masonry, and the arrangement of noncombustible stairs delayed the spread of the fire much more than an open stairway would have done.

The smoke of the fire was discovered by a bellboy who was called to a room at 3:15 A.M. For some reason, however, the fire department was not called until 3:42 A.M., and, in the meantime, the bellboy was imprisoned by the dense smoke in the room to which he had been called.

The fire spread from floor to floor by ignition of the trim through the stairway. The dense smoke, even with a small amount of fire, made the stairs and corridors impenetrable. The wood doors and transoms were burned out from the fourth through the thirteenth floors. Where the transoms were closed, many of the rooms showed little damage. However, even in some of the undamaged rooms, inhabitants were suffocated. In one room on the top floor, five women died from asphyxiation although the fire did not get into the room at all.

In all probability, the spread of the fire could have been completely prevented by the enclosure of the stairway. Undoubtedly no lives would have been lost if this enclosure had been accompanied by the use of some type of steel doors and if transoms to the rooms had been eliminated.

Our Lady of the Angels School—Chicago, Ill.

This fire is included here not because it is unusual or carries with it any lesson we could not have known already, but rather because it is so typical. Change the name of the place it occurred in, the number of casualties, and some of the details of construction, and the story of the fire which destroyed Our Lady of the Angels School [165] in Chicago, Ill., December 1, 1958, could have been the story of dozens of other fires which have claimed the lives of so many children in the United States and Canada in recent years. In this instance, 93 people lost their lives through failure on somebody's part to observe the known fundamentals of fire safety.

The building was erected in 1910, remodeled in 1951, and inspected by city officials just two months before the holocaust. It was a U-shaped building with masonry bearing walls and wood interior construction, two stories and basement in height. Interior partitions were of wood studs with ordinary lath and plaster. Each wing was approximately 60 ft wide by 120 ft long, and each wing contained three wooden stairwells.

Apparently the fire started in the basement near one of the stairways, smouldered for some time, probably in an accumulation of paper and rubbish in the basement under the stairs, then worked itself up the stairs to the first and second floors. A pair of steel doors at the first floor level protected that floor from damage, but the second floor level had no such protection, only an open archway to the corridor. It is difficult to understand how the fire spread so rapidly through the second floor without some kind of warning. Apparently the panic that accompanies such a disaster was the main cause of the deaths, many of which were the result of suffocation or superheated fumes.

As invariably happens when a fire in a school, hospital, or home for the aged takes a toll of lives, immediate investigations of fire hazards were conducted in schools throughout the country. The *Engineering News-Record* listed the results of the investigations in some of the cities as follows:

(1) Philadelphia, Pa., found that 96 per cent of the schools were violating the fire code or tolerating unsafe conditions; (2) Detroit, Mich., reported 57 schools that had been built before 1912 when

standards of fire-resistant construction were established there; (3) The Kentucky state fire marshal pointed to the fact that the fire described could happen in 75 per cent of that state's schools; and (4) New York, N.Y., closed 36 schools, pending correction of violations of the fire code.

It would be interesting and informative if a resurvey of the various city and state reports could be made, to show how many of these hazardous conditions were corrected as a result of the surveys made after the Chicago school fire.

General Motors Corp.—Livonia, Mich.

The worst industrial plant fire in history destroyed the General Motors Hydromatic automatic transmission plant [166] at Livonia, Mich., August 12, 1953, resulting in $55 million damage, killing three men, and demonstrating again that nonflammable construction isn't always fireproof.

The building, supposedly the last word in construction for automobile factory use, had been built four years earlier, a one-story building with a small second story section at one end. The framing was of unprotected steel with columns spaced 40 and 48 ft on centers. Above a low, brick apron wall, the exterior walls were made up mostly of steel sash. The roof was a flat, light-gauge metal deck, covered with mopped-on asphalt and felt, then a ¾-in.-thick layer of Fiberglas, over which were more layers of asphalt and felt, with tar and gravel surfacing.

The fire started when sparks from an oxyacetylene cutting torch ignited oil in a conveyor drip pan 11 ft above the floor. The drip pan caught the drippings of a rust-inhibiting liquid with a flash point of about 100°F from parts being conveyed from a dip tank to the next operation. An attempt was made to control the fire with hand extinguishers, but it spread the full 120-ft length of the drip pan, warped the pan, and spilled over onto the wood-block floor. Oily condensate on the underside of the roof deck also caught fire, and warping of the metal deck permitted the asphalt to drip through and ignite. An interval of from 6 to 20 minutes had elapsed before the Livonia Fire Department was called, the length of time being in dispute. By the time the fire company was able to get a hose on the fire, it had spread to such an extent that the stream could only reach the outer 50 to 75 ft of the 866-ft width of the building. As a result, the fire could only be left to burn itself out the next day.

The story can best be told by a direct quotation from the October 1953 issue of the *National Fire Prevention Association Quarterly* in which blame is placed on:

1. An undivided fire area of 1,502,500 square feet (34.5 acres) in which absence of fire walls and roof vents denied access for fire fighting and prevented localization of heat and smoke.

2. Inadequate sprinkler protection (only 20 per cent of the total area protected—no sprinklers where fire started).

3. Incompletely engineered processes. Fire protection for the dip tank did not protect the drip pan (under a conveyor immersing parts in oil in the dip tank). Due to the processes, oily deposits had a tendency to accumulate and increase the fire hazard of several heat-treating areas; yet this condition does not seem to have had any engineering attention.

4. Unprotected steel construction, in particular the thin steel deck that did not offer sufficient insulation between banking heat and the built-up roof covering to protect asphalt from melting and dripping through the joints of the heat-warped deck. Steel trusses collapsed in a matter of minutes.

5. Use of an oxyacetylene torch under unsafe conditions.

6. Lack of an effective fire brigade.

7. Delayed fire department notification.

Sherry-Netherland Hotel scaffolding—New York, N.Y.

On the evening of April 12, 1927, a spectacular fire attracted much publicity when the scaffolding of the tower of the Sherry-Netherland Hotel[167] in New York, N.Y., caught fire from some unknown source. The structural damage was comparatively light; the total damage was estimated at $25,000, largely in the scaffolding itself. The scaffolding extended from the thirty-eighth story to the top of the tower and although the standpipe was in place with a Siamese connection the cap had not been placed on it, and the firemen were unable to use it. An attempt was made to battle the fire from the standpipe at the top of the Savoy Hotel across the street, but this proved ineffectual, and the firemen had to content themselves with saving the construction hoist from the twentieth story down, the maximum height to which the water would reach.

Aside from the damage to the scaffolding and hoist tower, minor structural damage to the permanent structure resulted. Some of the exposed steel framework of the tower was distorted, three elevator motors were ruined, and some spalling of concrete fireproofing took place where formwork had been burned.

Similar scaffolding or formwork fires occurred at the Riverside Church in New York City in 1928, and at the Foreman National Bank Building in Chicago in 1929.

10.3 Explosions

The subject of explosions has a certain amount of interest attached to the simple story of their effect on buildings. It has therefore seemed advisable to include here a few examples of this type of failure.

Continental Grain Elevator—Buffalo, N.Y.

Is it possible to have an explosion without any chemical reaction? This was the question put to the author in the investigation of a grain elevator failure in 1954. The Continental Grain Elevator was a 36-year-old reinforced-concrete elevator consisting of the usual battery of silos, 21 ft 6 in. in diameter, with interstices. One of the interstices had been filled with a load of wet corn for drying. Samples from the bottom of the bin indicated that the corn had dried sufficiently, and they began drawing it out from the bottom of the bin. During the process, November 27, 1954, a piece of concrete from the face of the interstice fell out, crushing the roof of an adjacent office building; the corn, flowing from the opening, smothered a man in the office building. The piece of concrete which fell out was about 11 ft wide, 30 ft high, from a depth of 12½ ft below the top of 43 ft below the top, and about 8 in. thick.

There seemed to be little argument as to what had happened. The corn below had dried out enough to flow freely, but a wedge of corn, swollen with moisture, had remained arched across the top of the bin. As the material was drawn from the bottom of the bin, a partial vacuum was formed until finally the combination of vacuum and weight of material caused the bond between the corn and the concrete to break, and the mass slid downward in the bin, acting as a plunger or piston to compress the air in the hollow portion of the bin. The compressed air, changing suddenly from a vacuum to an outward pressure, built up sufficient force to blow out the side of the bin.

The only insurance coverage the owner had against anything like this kind of an accident was explosion insurance, and the insurance adjusters claimed that, since an explosion involved a chemical change, this accident was not an explosion. The author was asked to give an opinion, and stated that an explosion had occurred—a compressed air explosion—such as might occur in a popgun. After some further discussion of the nature and attributes of an explosion, the owner received his insurance payment.

Consolidated Elevator Co.—Fort William, Ont.

This failure, similar in nature to the preceding one, occurred at Fort William, Ont., at the Consolidated Elevator Co.,[168] October 29, 1910 (Fig. 10.7–10.9). The silo involved in the failure was of tile and reinforced concrete, and for the purpose of making the elevator more elastic in its use, a steel bin with a hopper bottom had been installed in the top third of the silo. Under a heavy load of wheat the steel hopper bottom tore loose, and, acting as a piston under the weight of the wheat, compressed the air in the middle third of the silo above the oats in the lower third, causing the front of the middle

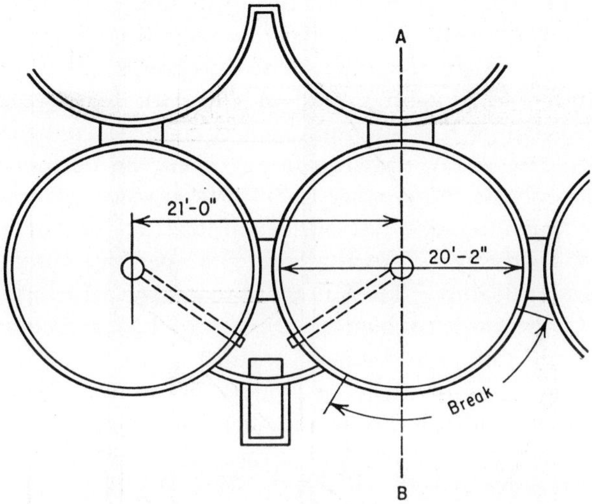

FIG. 10.7. Partial plan—Consolidated Elevator Co. silos

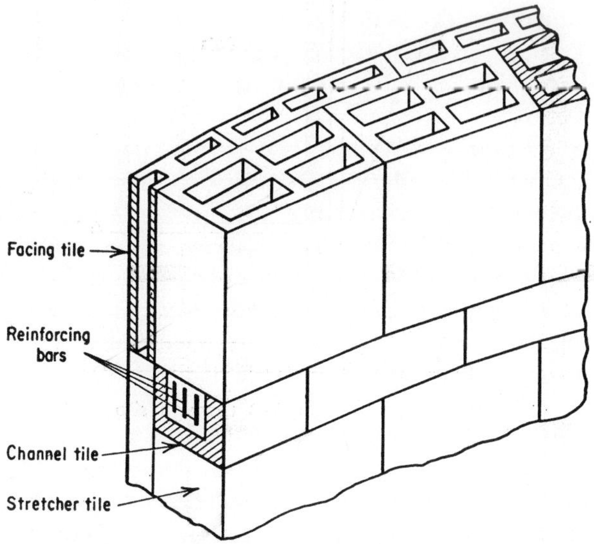

FIG. 10.8. Isometric view of bin wall detail

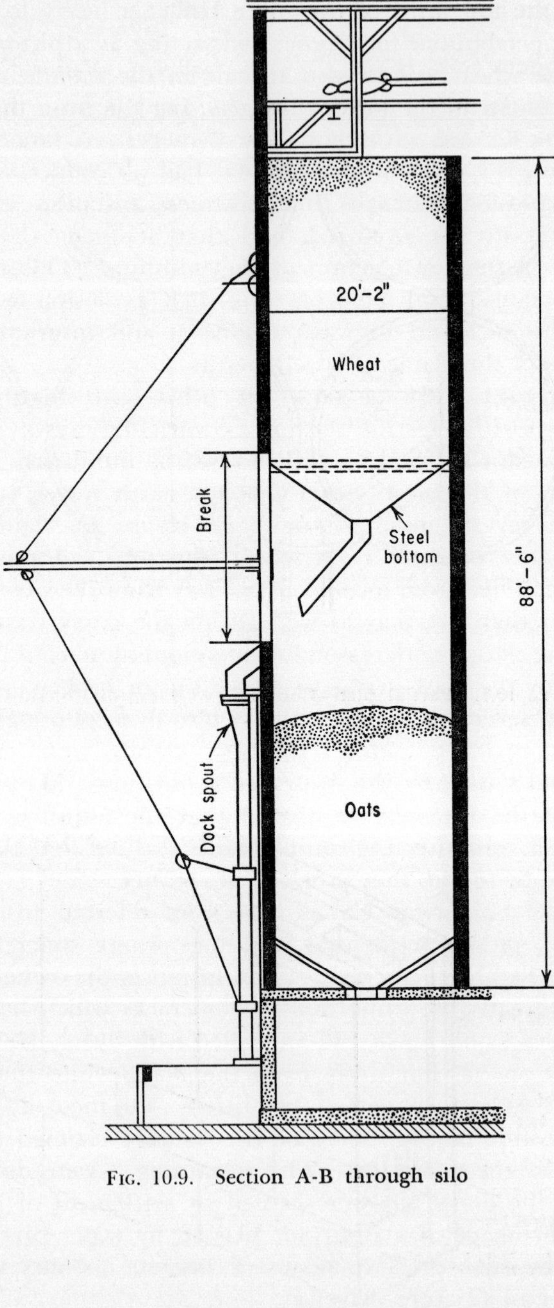

Fig. 10.9. Section A-B through silo

third of the silo to blow out, spilling the grain on the deck of a boat alongside. No injuries or deaths occurred.

Tank car—Niagara Falls, N.Y.

Whenever a serious explosion of any kind occurs, the insurance companies are flooded with claims for damages. A large proportion of these claims is based on damages that had no connection whatever with the explosion—damages that in most instances existed long before the catastrophe occurred, and that are generally caused by minor faults in the construction of the building. This brief discussion of the claims resulting from one such explosion will serve to emphasize this fact and to warn engineers and insurance adjusters of the nature of the claims that will result.

On January 21, 1958, a tank car filled with highly explosive chemical let go with a terrific blast on a connecting railway in a remote section of Niagara Falls, N.Y., with no nearby buildings.

The results of the blast, either through earth waves or air shock, were felt for several miles around, and claims for damages began to pour in—most of them from within this area, although some of the claims were filed from owners miles away from the site—apparently on the theory that "we may as well get on the gravy train."

Many of the claims were reasonable in amount and could easily have been caused by the blast—windows blown out or broken—walls moved or cracked, in several instances actually blown in, although no building collapse occurred. Most of these claims were immediately recognized and settled by the insurance companies. Many at the opposite end of the list were so absurd that the adjusters refused to recognize them, and after the complainants realized that there was no easy money to be found, they dropped the matter.

This left 20 borderline claims that were referred to the author to investigate, mostly residential buildings where concrete masonry walls or floors had been cracked. In all instances the owner magnified the damages greatly, pointing out plaster cracks which he attributed to the blast, cracks obviously old and filled with dust. He pointed to cracked walls on the side away from the blast, while the long wall toward the blast was in perfect condition. Of these 20 claims, 12 had some possible merit. Some of the damage claimed could have been caused by the explosion. The remaining 8 were definitely not the result of the blast, but were caused by settlement of part of the footing, by shrinkage of masonry or plaster, by water pressure under the floor, or by earth pressure against a concrete masonry wall caused by the load of an adjacent driveway.

Ohio State Office Building—Columbus, Ohio

A gas explosion that caused more than $1 million damage to the almost completed Ohio State Office Building [169] at Columbus, Ohio, and the death of 9 men and injury to 53 others on April 14, 1932, is typical of many smaller blasts that have occurred at various times.

The gas had apparently leaked from a 16-in. gas main about 40 in. under the sidewalk at a point about 15 ft from the wall of the building. The leak was 11 in. long and 1/20-in. wide. Apparently the gas found its way to a pocket under the subbasement floor of the building, where it was somehow ignited.

In the explosion that resulted, the twelve-story steel frame was uninjured. The concrete joist floor above the subbasement was lifted off the beams and dropped into the subbasement, and some damage was done to the concrete walls. The 70 × 160-ft slab in the first floor over the subbasement area was also lifted off the structural steel frame and dropped into the subbasement. The shafts of two banks of four elevators each, near the center of the explosion, carried the pressure up, and it diminished as it went, but it caused a great deal of damage to partitions throughout the building and blew out the windows of the two lower floors, including the frames. The concrete fireproofing of the beams from which the floor slabs were blown off remained intact, stripping the concrete in some instances from the overlapping bars of the concrete joist floors. The fact that damage was done to much of the finish—the plaster, bronze, and other ornamental work—greatly increased the money value of the damage done.

New London High School—New London, Tex.

One of the most costly building accidents in terms of human lives the world has ever known occurred March 18, 1937, when an explosion in a school [170] at New London, Tex., killed 455 persons. The school was a steel frame building with concrete floors. It was an E-shaped building, with a center stem three stories high and other stems one and two stories, respectively.

The building was heated with individual gas-fired radiators in each room. The gas had formerly been bought from the local utility company, but, in January, two months before the explosion, it had been changed over to use a waste gas from a gasoline stripping plant. In some way the gas, which is odorless, escaped into the basement below the three-story wing, and an ignition spark set off the explosion that completely wrecked this part of the building and killed so many of its occupants.

11

Minor and Incipient Failures

Although the cost of some of these failures cannot be considered minor, the reason for using this term is that no complete collapse is involved, and in some of the cases—called incipient failures—a collapse was averted, even though it cost real money to save the structure.

11.1 Minor Failures

These failures, though comparatively inexpensive and nondestructive, are of interest because they point up widely differing reasons for buildings failing.

Curling of industrial plant floor slab

The floor slab for an industrial plant floor was poured late in 1956 for very precise manufacturing processes. It was a one-story building with an area of about 120 × 240 ft and the floor was laid in a checkerboard pattern with 30-ft squares to minimize shrinkage difficulties, with construction joints but no expansion joints.

A well-tamped fill was placed under the floor—up to 2 ft at one end. The specifications originally called for foundry-sand fill, but because of the difficulty in getting it, fly ash was substituted.

As the slabs cured and dried out, many of the 30-ft-square slabs curled along the edges, lifting from the fill, and an electric lift truck traveling over the joint caused the slab to rock to such an extent that the precise work of the machinery was disturbed. At the corners of the slab, this curling was accentuated by lifting in both directions. The edge of the slabs curled from 1/4 to 1/2 in., and the corners curled as much as 3/4 in.

The curling probably had two contributing causes: the more rapid evaporation along the edges of the squares than in the center, and the absorption of moisture from the concrete by the hygroscopic fly ash fill.

The condition was cured and the floor stabilized by drilling the floor near the edges and grouting under the slab, then grinding down the top of the floor where it had lifted.

Wurlitzer Co.—North Tonawanda, N.Y.

In 1923 an office and factory building was built for the Wurlitzer Co. in North Tonawanda, N.Y. Most of the building was three- or four-story reinforced-concrete construction, carried on spread footings on a layer of sandy clay 3 to 4 ft thick, overlying 6 ft of medium hard clay, which was over 20 ft of soft clay and 3 ft of hardpan on rock (at about 35 ft below grade).

In the center of the plant, over the office portion of the building, a large water tank stood; the insurance company insisted that the tank be carried on piles to rock.

Until 1946 the building functioned without any signs of distress. Then a differential settlement appeared between the steel framework carried on the pile foundations and the concrete framework on spread footings surrounding the steel portion on three sides. The settlement was gradual, but the bay immediately adjacent to the pile-supported structure showed a definite drop. In this one bay, cracks began to appear in partitions, doors were bound, and serious cracking appeared in concrete beams. Levels were taken, and they indicated a drop of almost 3 in. in one bay with a gradual rise through the four bays beyond until the level beyond this point was practically the same as the level within the limits of the steel portion of the building.

Although some vibration could be detected, it was apparently no more serious in the area where the maximum settlement had occurred than in the bays beyond the limits of the settlement. The equipment used on the upper floors was relatively light and smooth running and was spread rather uniformly over the entire floor area.

Two small air compressors on the ground floor were suspected, and, at the suggestion of the investigating engineer, the larger of the two—a single-stage compressor—was shut down. The settlement continued at about the same rate, and, in an attempt to determine the cause of it, an Ames gage was applied to the side of a steel beam. It showed that when the smaller compressor was operating, pulsations up to

0.006 in. were registered, and when the compressor cut out, the pulsations dropped to 0.001 in. When the number of pulsations per minute were counted, they were found to coincide with alternate pulsations of the two-stage air compressor. The compressor was moved to another location, and when the floor was dug up it was found that the foundation for the compressor had been placed directly on top of the concrete of one of the building foundations. After the compressor was moved all appreciable settlement of the building stopped.

Philadelphia General Hospital Food Service Building—Philadelphia, Pa.

When politicians make decisions on technical matters for political purposes, the probability is that trouble will ensue. The blame is usually placed on the architect or engineer, who had the right answer the first time, but whom the politicians overruled. This is what happened in connection with the troubles that developed in the $3.4 million Food Service Building [171] of the Philadelphia General Hospital, in Philadelphia, Pa.

The building had been designed by a veteran Philadelphia architect of excellent reputation before his death in 1951, as a five-story brick, glass, and aluminum-faced structure, but, with the change of political color of the administration, it became a political football. Democrats claimed it was overdesigned and called it "a doctor's palace." It was partially redesigned, substituting cheaper construction for the glass and aluminum but, in general, holding to the original concept.

Before the building was completed, cracks began to appear in the partitions and in the exterior wall facing. Then the arguments and buck passing began. Two inspections were made—one by the newly appointed architect and the other by a firm of consulting engineers. They arrived at the same answer. The cracks did not affect the safety of the building and were, no doubt, the result of omitting expansion joints, except in the top story. Whether this was the cause of the cracks or not was disputed by other Philadelphia architects.

If this was the cause of the cracking, the responsibility was still in doubt. The contractor had requested that additional joints be used, and the supervising architect concurred in the decision. The economy-minded city officials, however, refused to authorize any extra work. In the meantime, a new city charter had been adopted, changing lines of responsibility, and there had also been a complete turnover of Public Health officials so that, regardless of who was responsible, it was necessary in 1957, for the city council to authorize the expenditure of $70,000 to make repairs in the building.

Corning Hospital partition cracks—Corning, N.Y.

In 1951, serious cracking occurred in the transverse partitions of the Corning Hospital in Corning, N.Y., which had been in use for two years. Level readings indicated that throughout the building, the exterior walls had settled about 3/8 in. more than the corridor columns. Horizontal shear cracks appeared in cross-partitions. The lower part of the partitions was, for the most part, terrazzo wainscot on cinder concrete block, and the upper part to the ceiling was plaster on gypsum blocks.

The shear crack occurred mostly along the line between the two partition materials. In cross-partitions, which were broken up by door openings, the cracks appeared in two places, the point mentioned and along the top of the base.

The foundations were in deep glacial gravel not far from the river, which usually rose considerably during spring thaws or after a particularly heavy rainfall, and tests indicated a large percentage of voids.

Where the permanent load of a wall-bearing brick wall was on this material, it had gradually compressed to a greater extent than had the gravel under the interior column footings where the full design load had never been reached because of the live load design required by the building code.

As a result, the differential settlement had occurred between the exterior and interior footings, causing the horizontal crackings of the cross-partitions.

Skating rink—Berkeley, Calif.

Settlement of foundations because of lowering of the water table has been discussed in several other sections of this book. A directly opposite effect will result from a "permafrost" condition in some cases. The worst case of this type that the author knows involved an indoor ice skating rink [172] in Berkeley, Calif., in 1956, where ice, which had penetrated the ground to a depth of 22 ft, raised the rink floor 3 ft.

Red Wing Packing Co.—Fredonia, N.Y.

The damage from frost heaving in a four-story, flat-slab, reinforced-concrete warehouse of the Red Wing Packing Co. at Fredonia, N.Y., in 1953 was on a much smaller scale than that described in the preceding article, but was potentially more hazardous. A portion of the building showed floor cracks, which indicated that the building generally was rising. Levels indicated a differential reading of approximately 4 in.

Part of the basement had been converted into a deep-freeze room, and the trouble began to develop shortly afterward. A 3-in.-diameter hole was drilled 6 ft deep next to an interior column footing and it was found that the ground had frozen beneath the footing.

After further investigation, the faces of the columns were wound spirally with electric heating wires and the columns and heating wires were encased with corkboard insulation. In less than eight months the columns had returned to their original position and floor cracks were closed up.

Leaning church steeples

This type of failure is so common that, instead of writing up any particular case, a composite of several of them will be sufficient, describing points that are common to more than one in almost every instance.

For the most part, these churches were built in the second half of the nineteenth century, when wood trusses were used to span auditoriums—especially in towns and villages where steel erection meant bringing erectors from the big city, but where wood trusses could be built by capable carpenters from materials easily obtained locally. The steeples themselves, in such cases, are practically always of wood framing. The author has never been called upon to inspect a steeple of solid masonry construction or with structural steel framing.

One of the primary causes, in many of these leaning steeples, was a rotting condition at the base of the steeple. Moisture entered, either as rain through the louvers in the steeples, or through insecure flashing, where the steeples joined the roof, and this moisture attacked the end grain of the wood posts of the steeple and decay began. Because of their relatively high ratio of height to width of base, wind action on the steeple brought about high stresses in alternate directions, a condition designed to tear apart the metal flashing by alternate bending. Often, because of the small space available within the steeple, bracing against the action of the wind was weak, and the movement of the steeple tended to loosen the nails, thus further weakening the structure.

In many of these steeples, the front posts were carried on the top of a masonry wall which was relatively rigid, whereas the back posts were carried either directly or indirectly by wood trusses, which changed in volume, and therefore in level, as they dried out through the years.

Because they are not easily accessible, the points of danger are inspected very infrequently, and, unless the spire obviously starts to lean or water stains appear on the ceiling under the steeple, the deteriorating condition is not suspected. Situations that would be guarded against in a private home by the owner or in a business building by the management, will often go unattended in a church, sometimes for lack of funds, sometimes merely because what is everybody's business is nobody's business.

Strange as it may seem, even though church bells in such a tower may cause vibration, they very seldom bring about this type of failure. Generally such bell towers are more accessible and the very existence of the bells causes them to receive more attention.

Growth of concrete

For a period of a few years during the middle of the fifties, the construction industry in some areas was plagued by the phenomenon of growth of concrete—a phenomenon diametrically opposite to what we had become used to—cracking due to shrinkage. During this

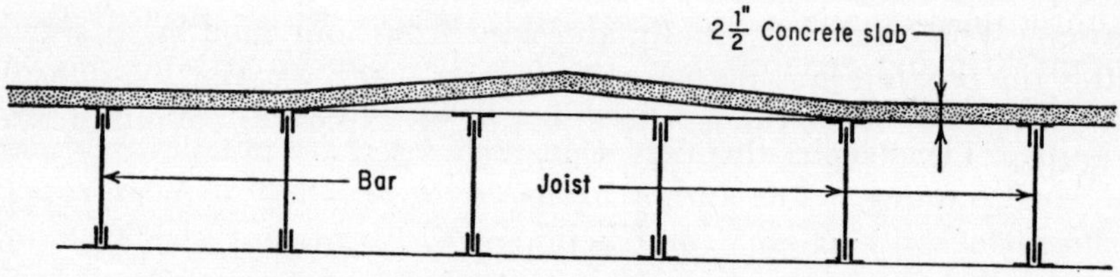

FIG. 11.1. Floor slab lifted from bar joist by concrete growth

period the author was called in on at least a dozen structures where this condition manifested itself. Although this condition still shows up occasionally, it is not nearly as frequent nor as severe as it was then.

The growth of concrete manifested itself in different ways on different jobs. On a suburban school job, a 2½-in.-thick concrete floor slab was lifted about 1½ in. off the steel joists on which it was bearing, by the growth of the slab (Fig. 11.1). On a one-story warehouse floor a track conveyor system used a moving chain in a steel box in the floor to pull small package trucks through the building. The growth of the floor slab put enough pressure on the upper portion of the steel box so that the mechanism that conveyed the truck would bind, and it became necessary to make saw cuts and put expansion joints in the concrete next to the tracks (Fig. 11.2).

In several places, this growth manifested itself by pushing out the corner of the concrete foundation wall and the brick wall (Fig. 11.3). If the question is asked, "Why didn't the concrete wall and the concrete floor both grow?," the answer is that they probably did.

However, the concrete foundation wall was placed some time before the concrete floor, and the wall had taken its growth before the floor was placed. In one place, a similar corner condition cracked a 4-in. thick precast-concrete facing stone placed at a similar corner.

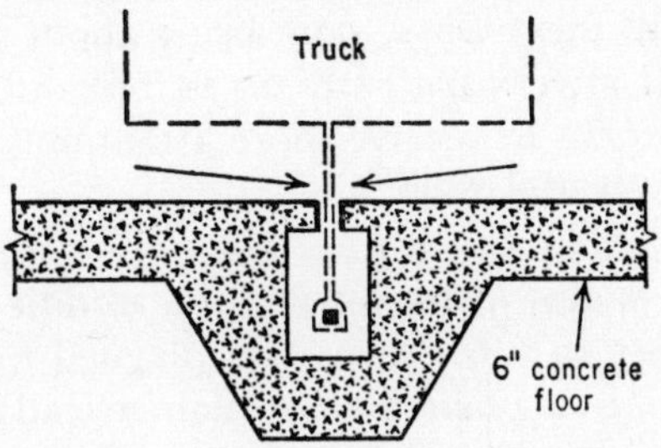

FIG. 11.2. Binding of conveyor chain caused by concrete growth

The above cases are only a few of those that occurred during the fifties. The condition usually appeared from four months to a year after the concrete was placed and, in most cases, was not too difficult to correct. It is mentioned here mainly to explain a condition that

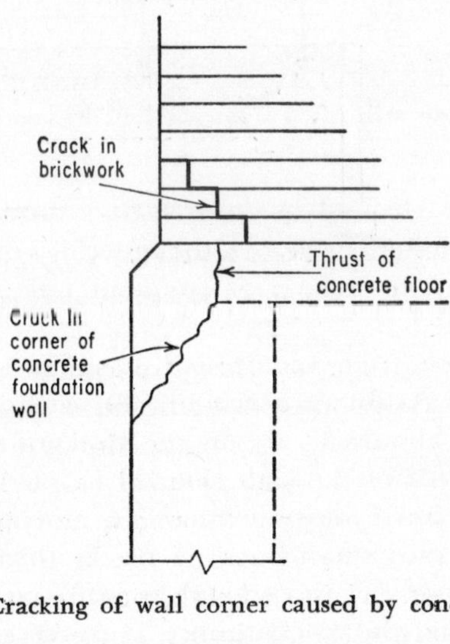

FIG. 11.3. Cracking of wall corner caused by concrete growth

may have been brought to the attention of other engineers, architects, or owners of buildings erected during that time.

The cause of this phenomenon is and has been the subject of much argument. The answer suggested here is only the theory of the author and is disputed by several competent concrete authorities. This theory is that the growth was caused by a chemical additive in the Portland cement, that had not been used prior to this time, and that has been omitted since that time. This theory was never accepted by some others who suggested that the element causing the growth was improper aggregate. However, in the cases brought to the author's attention, the cement appears to have all been from the same source, whereas the aggregate was from half a dozen sources, quite remote from each other.

Concrete masonry basement walls

As has been noted in Sec. 6.1, in ordinary house construction and often in the construction of store buildings or other similar structures,

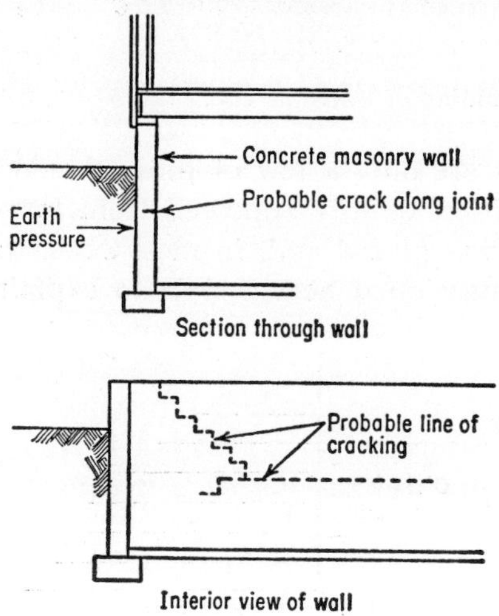

FIG. 11.4. Failure of concrete masonry walls

concrete masonry walls—or as they are commonly called, concrete block walls—are used for enclosing the basement. Often, within a year or two, cracks start to open up along one or more of the horizontal joints at a point from one-third to one-half the depth from the ground level to the basement floor, resulting in leakage and, in a few instances, in complete failure of the wall. Often these cracks fade out toward the end walls or break roughly on a diagonal line following vertical and horizontal joints toward the top of the wall where the long wall joins the end wall (Fig. 11.4).

In Sec. 1.4, the effect of bending is discussed, and Fig. 1.2 and 1.3 indicate the resultant stress created by the bending. Earth pressure on the outside of a wall creates a vertical load on the layer below, but it also creates a lateral pressure on the wall and, therefore, exerts a bending action on the wall similar to beam action. The resulting stress is a compressive stress on the outer face and a tensile stress on the inner face of the concrete masonry. In addition to the bending stress, the weight on top of the wall exerts a compressive stress over the whole area. The result may be a compressive stress over the entire area, or a compressive stress over the outside face and a net tension on the inside face. Mortar has a very low tensile strength, particularly at the surface of the concrete masonry where the block may absorb most of the moisture from the mortar. The result, therefore, may be a failure in tension of the inside face of the concrete masonry wall, resulting in the crack of the wall as described.

A check on theoretical stresses indicates that with a 7-ft height from basement floor to underside of joists, an earth fill to a point 5 ft above the basement floor could be sufficient to cause such cracking. With the superimposed load of an adjacent driveway or a wet clay subsoil, cracking could result from even a lesser backfill.

Movement or tipping of a concrete wall during construction

In the construction of a poured-concrete basement wall, during the time before the floor construction is in place, the wall is subject to displacement or tipping if any load is placed against it. It becomes a retaining wall without being designed to resist retaining wall stresses.

The author has frequently been called upon to examine such a wall where the contractor has placed a backfill against it—generally using a heavy bulldozer. Sometimes the wall has been moved inward as much as 4 to 6 in. at the center of a long bow, and in others it has tipped out of plumb, usually held together by the reinforcing bars in the concrete, although sometimes serious cracking has occurred.

As a result of such movement or tipping of the wall, it has been the author's practice to specify that no backfill be placed until the basement and ground floors have been poured, thus providing sufficient support for the wall.

11.2 Incipient Failures

Incipient failures are important because they can serve to warn professionals in the building trade of possible pitfalls on future construction projects, in addition to salvaging the existing structure.

A potential "concrete floor arch" failure

A potential failure, rather than an incipient failure, would be a better way of describing this little incident which occurred during the PWA construction program of the thirties. It tends to point out that "a little knowledge is a dangerous thing" and that the improper use of words can have serious consequences, especially when these words have come to have a different meaning through the years.

The job was a simple, two-story suburban school, with exterior bearing walls, steel-beam-and-column framework, and reinforced-concrete blanket-type slabs, 4 in. thick and spanning from 7 to $7\frac{1}{2}$ ft. The clerk-of-the-works, representing the school board, was a superannuated building mechanic whose claim for the responsibility of the job was based on political influences rather than technical knowledge.

The structural designer routinely inspected the job to check the formwork and the placing of steel before the concrete floor was first poured. He was shocked to see that all of the bent bars were placed with the steel in the top of the slab at the center of the slab, and the lap over the beam at the bottom of the slab. He called the inspector's attention to the error and, in the usual direct, perhaps profane, manner wanted to know why the detail drawing for placing the bent bars had not been followed. In a correspondingly direct and profane manner, the inspector replied that it was because the detail drawing was all wrong—that any kind of fool would know that in a concrete arch, the steel had to be in the top in the center of the span, and weren't these floor arches? The designer explained that they weren't floor arches but slabs. In the early days of concrete construction, floor arches were cast from beam to beam, and this has led to a continued use of the improper designation—so that even today, more than half a century after the use of these arches has been discontinued, concrete floor slabs are still sometimes called "concrete floor arches" instead of slabs.

School truss—Detroit, Mich.

The *Engineering News-Record* [173] describes an incipient failure that could have been far more serious, except for an alert superintendent of buildings for the school department.

The original design of a 60-ft steel truss over a gymnasium called for the chords to be a full 60-ft length without any splice. Because of erection conditions, the lower chord was spliced at the end panel point. The truss was encased in a plaster partition, and, when a crack developed in this partition, the superintendent became suspicious, and removed the plaster. He found that the joint had failed completely, but in spite of the failure nothing had collapsed. The gusset plate had been the only splice for the member; no bottom splice plate or angles were used (Fig. 11.5). The truss, which had been in service for sixteen years before the failure was discovered, had settled about 7 in.

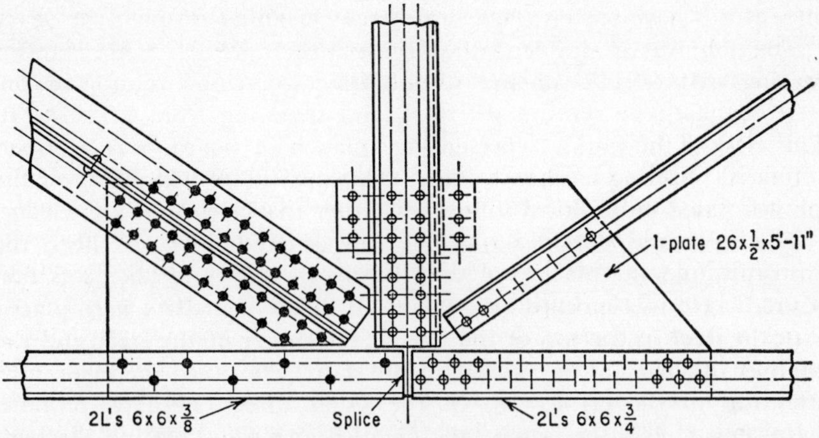

FIG. 11.5. Detail of weakened truss—Detroit school

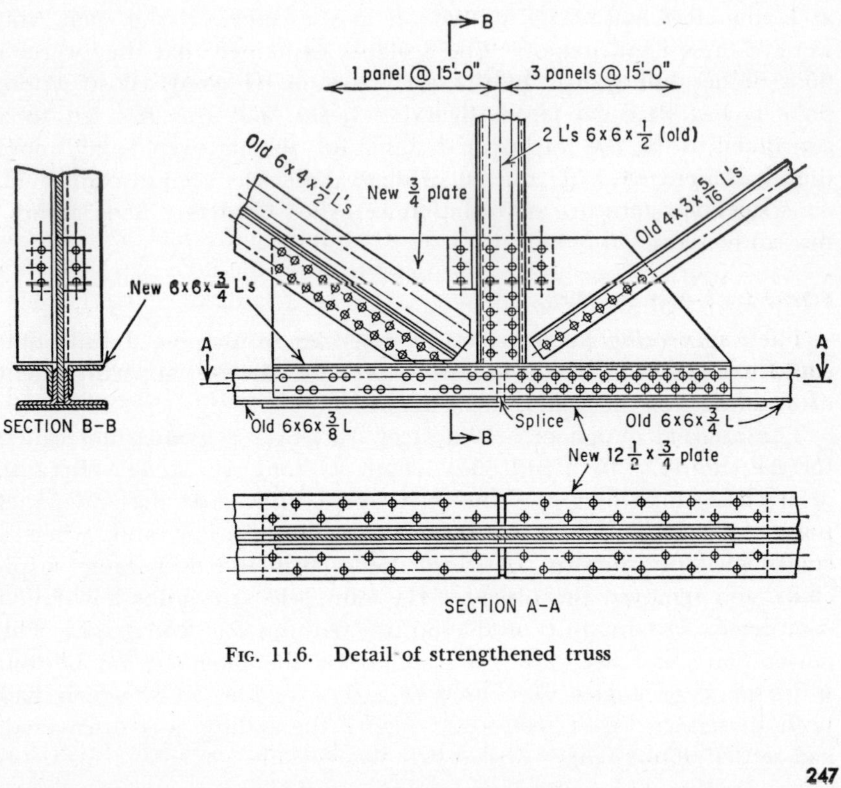

FIG. 11.6. Detail of strengthened truss

The truss was jacked up in place, a heavier gusset was put in, as well as a bottom splice plate and splice angle (Fig. 11.6). Although the opposite end of the truss had the same faulty detail, it showed no signs of distress. The repairs described above were made, and the structure was put back into use again.

Kindel Bed Co.—Long Island City, N.Y.

This is the story of a failure that should have happened, but did not. In 1916, the Kindel Bed Co. in Long Island City, N.Y., built a four-story reinforced-concrete building,[174] 80 ft wide and 218 ft long. The building was three bays wide with columns spaced 18 ft on centers in the length of the building and beam-and-girder construction with beams 6 ft on centers, three to an 18-ft bay. It was designed for a 200-lb live load. The original owners vacated the building, and the second floor was leased for storage of boxes and sheet tin for the Standard Oil Co., each box $14 \times 8\frac{3}{4} \times 4$ in., weighing 122 lb. The entire floor, with the exception of a 4-ft aisle, was piled as high as twenty deep with these boxes.

One night, early in 1919, the night watchman heard rumblings and crunchings and, in the interest of his own safety, he left the building and stayed out the rest of the night. The next day he reported the trouble and, upon examination, the floor was found to be badly cracked and in some places deflected. It was found that in some places the loads were as much as 1,300 psf, and the entire floor had the equivalent of 900 psf, four-and-one-half times its design load. This does not mean, however, that the concrete was overstressed to four-and-one-half times its load, or that it actually had a factor of safety of four-and-one-half. The dead load on the floor was probably 100 psf, so that the original design was on a basis of 300 psf. The actual load was 900-psf live load plus 100 psf dead load, or 1,000 psf, still three-and-one-third times what it was designed for and far in excess of the factor of safety of two, which would normally be expected.

Bowling alley—Buffalo, N.Y.

In October 1958, the author was called by a real estate agent to inspect the second floor of a two-story building where the tenant complained of excessive vibration. The structure was probably 30 years old, of wood-joist, steel-beam, and pipe-column construction. The ground floor was occupied by stores, and the second floor contained bowling alleys and a bar. The area in which vibration might be expected—the bowling alley—was perfectly stable, but the floor under the bar was quite shaky.

Investigation disclosed that when the bar was installed, to permit access to the plumbing under the bar, six of the existing joists were cut and reversed in direction without doubling or otherwise reinforcing the header joist. As a result, the two joists to which the new joists were attached, were carrying the floor load originally carried

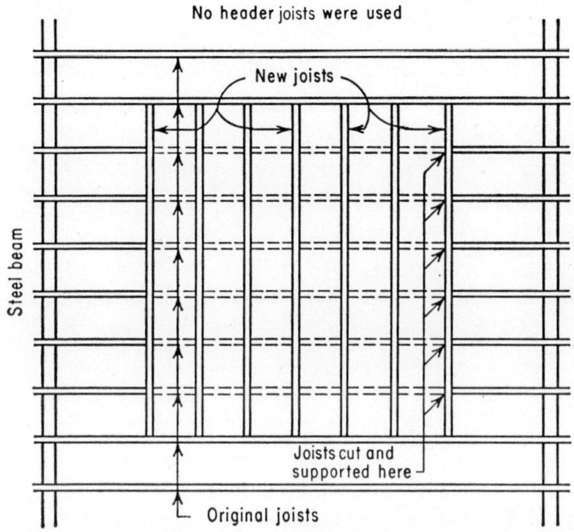

FIG. 11.7. Potential joist failure of bowling alley due to building alterations

by seven joists, or approximately three-and-one-half times their original load (Fig. 11.7).

A package deal in the delivery of reinforcing steel

Some years ago a contractor whose specialty is concrete repair work was called in to gunite or otherwise repair the soffits of a series of reinforced-concrete beams where horizontal cracks were developing along the plane of the bottom reinforcing steel.

When the loose soffits of the beams were removed, it was discovered

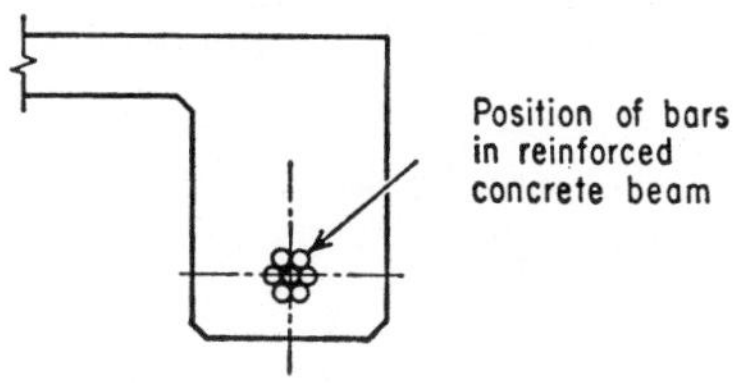

FIG. 11.8. Improper placing of reinforcing bars

that the bent bars had been dropped into the beam forms, all wired together neatly in bundles with metal tags containing the beam numbers, just as they had been received on the job (Fig. 11.8). The building was upwards of 20 years old, and how the soffits had remained on the bars as long as they did is anybody's guess.

References

CODE OF ABBREVIATIONS

A.C.I. American Concrete Institute
A.F. Architectural Forum
A.R. Architectural Record
B.A.I.A. Bulletin of the American Institute of Architecture
C.E. Civil Engineering
C. Eng. Construction Engineering
E.E. European Engineering
E.N. Engineering News
E.N.R. Engineering News-Record
E.R. Engineering Record
J.A.I.A. Journal of the American Institute of Architecture
N.F.P.A. National Fire Protection Association Quarterly

Chapter 2

1. A.C.I., Dec. 1957, p. 449
2. E.N.R., May 12, 1955, p. 23; July 14, 1955, p. 23; C. Eng., June 1955 p. 84; July 1955, p. 33
3. E.N.R., June 23, 1949, p. 36
4. E.N.R., Oct. 28, 1937, p. 1619
5. E.N.R., Apr. 16, 1931, p. 666
6. E.N.R., July 26, 1934, p. 125
7. E.N.R., Apr. 21, 1949, p. 13; A.C.I., Dec. 1949, p. 467
8. E.N.R., May 20, 1937, p. 759
9. E.N.R., Aug. 8, 1942, p. 55; A.C.I., Dec. 1957, p. 467
10. E.N.R., July 28, 1921, p. 165; Sept. 15, 1921, p. 431
11. E.N.R., Nov. 19, 1959, p. 29
12. E.N.R., Mar. 17, 1960, p. 23
13. E.N.R., June 23, 1960, p. 42

15. E.N.R., Oct. 10, 1940, p. 449
16. E.N., Apr. 14, 1910, p. 447
17. E.N., Oct. 19, 1911, p. 489
18. E.N., Feb. 10, 1916, p. 294
19. E.N., Dec. 26, 1912, p. 1233
20. E.N., Nov. 15, 1906, p. 517; Nov. 29, 1906, pp. 555–573; Dec. 6, 1906, p. 598; Dec. 20, 1906, p. 661
21. E.N., Jan. 3, 1907, p. 1; Jan. 31, 1907, p. 130
22. E.N., Nov. 27, 1913, p. 1104; Dec. 18, 1913, p. 1254; Jan. 8, 1914, p. 96; Hammond, Rolf, *Engineering Structural Failures,* Philosophical Library, N.Y.
23. E.N.R., Feb. 7, 1924, p. 239; Apr. 3, 1924, p. 556; A.C.I., Dec. 1957, p. 458
24. E.N.R., Nov. 12, 1925, p. 800
25. E.N., Feb. 29, 1912, p. 415; Mar. 14, 1912, p. 499; Mar. 28, 1912, p. 601
26. E.N., Apr. 5, 1906, p. 396; A.C.I., Dec. 1957, p. 458
27. E.N., Jan. 7, 1904, p. 21; Jan. 28, 1904, p. 82; A.C.I., Dec. 1957, p. 457
28. E.N., Jan. 9, 1913, p. 86
29. E.N., Dec. 3, 1914, p. 1142
30. E.N., Nov. 24, 1910, p. 582; Dec. 8, 1910, p. 686; Jan. 26, 1911, pp. 105–117; Feb. 23, 1911, pp. 231, 236; Mar. 2, 1911, p. 269; Mar. 16, 1911, p. 329
31. E.N., Apr. 9, 1903, p. 308
32. E.N.R., Feb. 8, 1923, p. 254
33. E.N.R., Dec. 3, 1942, p. 755; Dec. 10, 1942, p. 711

Chapter 3

34. E.N.R., Feb. 17, 1921, p. 294; Mar. 17, 1921, p. 479
35. E.N., Feb. 15, 1917, p. 293
36. E.N.R., May 14, 1959, p. 29; May 21, 1959, p. 128
37. E.N.R., July 1, 1937, p. 2
38. E.N.R., Dec. 8, 1921, p. 954; Jan. 5, 1922, p. 33; Feb. 2, 1922, p. 179; Mar. 23, 1922, p. 504
39. E.N.R., Dec. 30, 1920, p. 1294
40. E.N.R., May 29, 1930, p. 909
41. E.N.R., Dec. 20, 1934, p. 804
42. E.N.R., May 25, 1939, p. 43
43. E.N., May 10, 1906, p. 530
44. E.N.R., July 23, 1959, p. 25
45. E.N., Aug. 27, 1914, p. 470
46. E.N.R., Dec. 8, 1927, p. 937
47. E.N., Nov. 18, 1915, p. 1005
48. E.N.R., Dec. 20, 1951, p. 26
49. E.N., Jan. 30, 1913, p. 234; Jan. 1, 1914, p. 203; Mar. 26, 1914, p. 689
50. E.N., July 6, 1911, p. 30; July 13, 1911, p. 56; Aug. 3, 1911, p. 273; Jan. 11, 1912, p. 83

Chapter 4

51. E.N.R., June 25, 1936, pp. 928, 930; July 2, 1936, p. 28; July 9, 1936, p. 67; July 30, 1936, pp. 175, 177; Apr. 8, 1937, pp. 530, 533; May 13, 1937, p. 720; July 15, 1937, p. 93

52. E.N., June 18, 1908, p. 678; June 25, 1908, p. 693
53. E.N., Oct. 1, 1896, p. 209; Oct. 29, 1896, p. 273
54. E.N.R., Apr. 24, 1924, p. 723; Hammond, *op. cit.,* p. 87
55. E.N., Oct. 31, 1912, p. 817
56. E.N., Feb. 25, 1914
57. E.N., Apr. 25, 1912, p. 815
58. E.N.R., May 27, 1954, p. 27
59. E.N., Apr. 28, 1912, p. 750

Chapter 5

60. E.N.R., Aug. 22, 1929, p. 300
61. E.N.R., Oct. 13, 1927, p. 610; Dec. 22, 1927, p. 1012
62. E.N., July 22, 1909, p. 107; July 29, 1909, p. 131; Aug. 5, 1909, p. 159
63. E.N.R., Dec. 9, 1920, p. 1155; Feb. 17, 1921, pp. 282, 314
64. E.N., Dec. 12, 1912, p. 1113

Chapter 6

65. Holiday, May 1959; Encyclopaedia Britannica
66. E.N., Sept. 7, 1905, p. 248; Sept. 28, 1905, p. 335; Sept. 2, 1905, p. 265
67. E.N.R., July 16, 1925, p. 115; Aug. 20, 1925, p. 319
68. E.N.R., May 9, 1957, p. 75; Aug. 8, 1957, p. 44
69. E.N., Sept. 19, 1912, p. 555
70. E.N.R., Feb. 15, 1934, p. 245
71. E.N.R., Dec. 20, 1920, p. 1111
72. E.N.R., Aug. 19, 1926, p. 316
73. E.N.R., Nov. 9, 1922, p. 811
74. E.N.R., Jan. 22, 1920, p. 179
75. E.N.R., Oct. 8, 1959, p. 26
76. E.N., June 11, 1908, p. 642; June 18, 1908, p. 673
77. E.N., Mar. 12, 1908, p. 267
78. E.N., Nov. 5, 1908, p. 488
79. E.N.R., Sept. 15, 1955, p. 50
80. E.N., Apr. 25, 1907, p. 458
81. E.N., Dec. 16, 1915, p. 1189; Aug. 27, 1914, p. 464
82. E.N.R., June 23, 1955, p. 27; July 21, 1955, p. 12; Jan. 5, 1956, p. 27
83. E.N., Feb. 8, 1912, p. 268
84. E.N.R., Feb. 13, 1958, p. 44
85. E.N.R., May 13, 1920, p. 945
86. E.N., Nov. 4, 1909, p. 501; Nov. 11, 1909, p. 531
87. E.N.R., Aug. 10, 1933, p. 159
88. E.N.R., Oct. 11, 1956, p. 33
89. E.N., Dec. 21, 1911, p. 755
90. E.N., Aug. 2, 1906, p. 129; Aug. 9, 1906, p. 152
91. E.N., Oct. 19, 1911, p. 489
92. E.N.R., June 28, 1928, p. 1020; Nov. 8, 1928, p. 692
93. E.N., Aug. 15, 1895, pp. 104, 111; Aug. 22, 1895, p. 127; Aug. 29, 1895, pp. 140, 142; Sept. 5, 1895, p. 145; Oct. 3, 1895, p. 209
94. C.E., June 1955, p. 356

Chapter 7

95. E.N.R., Feb. 2, 1922, p. 179; Feb. 9, 1922, pp. 207, 221, 224; Feb. 16, 1922, pp. 264, 294; Feb. 23, 1922, p. 334; Mar. 23, 1922, p. 502; Mar. 30, 1922, p. 511; Apr. 6, 1922, p. 582; Apr. 25, 1922, p. 886; Aug. 9, 1926, p. 291; Sept. 2, 1926, p. 386
96. E.N., Mar. 10, 1904, p. 217; Mar. 17, 1904, pp. 250, 256; Mar. 24, 1904, pp. 279, 281; May 12, 1904, p. 454
97. E.N.R., Feb. 27, 1930, p. 377; Mar. 27, 1930, p. 539; Dec. 3, 1931, pp. 871, 899; Jan. 21, 1932, p. 105; Jan. 18, 1934, p. 199; Aug. 9, 1934, p. 191
98. E.N.R., Apr. 25, 1946, p. 5
99. E.N.R., Oct. 16, 1947, p. 75; Oct. 23, 1947, p. 2
100. E.N., June 22, 1916, p. 1184
101. E.N.R., May 10, 1928, p. 742
102. E.N., Mar. 23, 1905, p. 325; Apr. 13, 1905, pp. 327, 391
103. E.N., Mar. 10, 1910, p. 290; Mar. 17, 1910, p. 317
104. E.N.R., Apr. 5, 1923, p. 640; Apr. 12, 1923, p. 675; Apr. 19, 1923, p. 712
105. E.N.R., Jan. 20, 1927, p. 121
106. E.N., Aug. 29, 1912, p. 395
107. E.N., Mar. 7, 1912, p. 462
108. E.N.R., Mar. 6, 1947, p. 57
109. E.N.R., Jan. 7, 1926, p. 35; Jan. 28, 1926, p. 174; May 6, 1926, p. 744
110. E.N.R., Apr. 29, 1948, p. 73
111. E.N., Dec. 19, 1912, p. 127
112. E.N.R., Apr. 12, 1917, p. 85
113. E.N.R., Sept. 2, 1937, pp. 375, 412
114. E.N., Jan. 5, 1911, p. 27
115. E.N., Jan. 4, 1906, p. 23
116. E.N., Mar. 8, 1917, p. 399
117. E.N.R., May 21, 1931, p. 868
118. E.N., May 29, 1913, p. 1145; June 19, 1913, p. 1288
119. E.N., Nov. 13, 1902, p. 389; Nov. 20, 1902, p. 420

Chapter 8

120. A.C.I., Dec. 1957, p. 464; A.C.I., Jan. 1957, pp. 625, 637; E.N.R., Aug. 25, 1955, p. 23; Nov. 10, 1955, p. 21; Nov. 17, 1955, p. 120; Dec. 1, 1955, p. 10; Jan. 12, 1956, p. 10; Mar. 1, 1956, p. 10; Mar. 8, 1956, p. 14; Mar. 22, 1956, p. 10; Apr. 5, 1956, p. 9; Sept. 27, 1956, p. 28; Oct. 18, 1956, p. 11; Nov. 1, 1956, p. 12; Nov. 22, 1956, p. 6; Dec. 13, 1956, p. 10; C. Eng., Feb. 1956, p. 4
121. A.C.I., Dec. 1957, p. 454; E.N.R., Nov. 11, 1954, p. 36; Dec. 23, 1954, p. 6; Dec. 30, 1954, p. 8; Jan. 13, 1955, p. 10; Feb. 17, 1955, p. 10; Mar. 3, 1955, p. 6; Mar. 24, 1955, p. 8
122. A.C.I., Dec. 1957, p. 462; E.N.R., Mar. 27, 1952, p. 25; Apr. 3, 1952, p. 29; May 22, 1952, p. 25; July 3, 1952, p. 10; Oct. 23, 1952, p. 10; A.F., Apr. 1952, p. 47
123. A.C.I., Dec. 1957, p. 462; E.N.R., Apr. 12, 1956, p. 17; Apr. 19, 1956, p. 25; May 3, 1956, p. 25
124. A.C.I., Dec. 1957, p. 462; E.N.R., July 29, 1954, p. 29
125. E.N., Dec. 14, 1911, p. 717; Dec. 28, 1911, pp. 778, 790; Jan. 11, 1912, p. 66

126. E.N.R., Dec. 11, 1924, p. 950
127. A.C.I., Dec. 1957, p. 459; E.N.R., Sept. 30, 1956, p. 27; Oct. 18, 1956, p. 27; Nov. 6, 1956, p. 25; July 11, 1957, p. 28; Nov. 28, 1957, p. 12; Jan. 16, 1958; May 8, 1958, p. 10; June 6, 1958, p. 12; July 10, 1958, p. 8; C. Eng., Dec. 1956, p. 82
128. A.C.I., Dec. 1957, p. 466; E.N.R., May 28, 1953, p. 25
129. E.N.R., Dec. 2, 1943, p. 53; Dec. 16, 1943, p. 69
130. E.N.R., Mar. 7, 1935; Apr. 18, 1935, p. 573
131. E.N.R., Sept. 11, 1958, p. 27; Sept. 18, 1958, p. 15; Oct. 30, 1958, p. 8; Dec. 25, 1958, p. 8; Jan. 8, 1959, p. 23; Feb. 26, 1959, p. 13; Mar. 26, 1959, p. 11

Chapter 9

132. E.N., Oct. 21, 1897, p. 257
133. E.N.R., Nov. 17, 1927, p. 805
134. E.N.R., Sept. 28, 1922, p. 535
135. E.N., Dec. 2, 1909, p. 620
136. E.N.R., Dec. 25, 1947, p. 853
137. E.N.R., May 25, 1922, p. 873
138. E.N.R., June 26, 1930, p. 1069
139. E.N., Aug. 27, 1914, p. 470
140. E.N.R., Aug. 18, 1921, p. 271; Aug. 25, 1921, p. 323
141. E.N., July 17, 1902, p. 41
142. E.N.R., Nov. 4, 1937, p. 731
143. E.N.R., July 5, 1928, p. 32
144. E.N.R., Jan. 4, 1945, p. 3
145. E.N.R., Mar. 17, 1921, p. 483
146. E.N., Aug. 8, 1912, p. 272
147. E.N., Mar. 9, 1911, p. 307
148. E.N., Feb. 15, 1914, p. 328
149. E.N., Aug. 20, 1914, p. 395
150. E.N.R., Nov. 26, 1942, p. 738
151. E.N., Aug. 29, 1891, p. 190
152. E.N., Feb. 10, 1910, p. 168
153. E.N.R., Dec. 17, 1925, p. 1008
154. E.N.R., Feb. 27, 1936, p. 332

Chapter 10

155. E.N.R., Oct. 6, 1960, p. 32
156. E.N.R., Oct. 14, 1926, p. 621; Dec. 29, 1927, p. 1050, Mar. 1, 1928, p. 351; Sept. 30, 1926, p. 554; Oct. 17, 1926, p. 586
157. E.N.R., Aug. 22, 1935, p. 274
158. E.N.R., Feb. 11, 1937, p. 215
159. E.N., Dec. 17, 1914, pp. 1225, 1233; Dec. 24, 1914, pp. 1272, 1279
160. E.N., Nov. 16, 1916, p. 956
161. E.N., Apr. 5, 1917, p. 17
162. E.N., Feb. 2, 1911, p. 121
163. E.N.R., June 13, 1946, p. 899; June 12, 1947, p. 108
164. E.N.R., Dec. 19, 1946, p. 842; June 12, 1947, p. 492; The American City, Jan. 1947, p. 95
165. E.N.R., Dec. 11, 1958, p. 22; Jan. 22, 1959, p. 21

166. E.N.R., Aug. 20, 1953, p. 58; Oct. 15, 1953, p. 24; N.F.P.A., Oct. 1953, p. 105; B.A.I.A., Sept.–Oct. 1953, p. 161
167. E.N.R., Apr. 21, 1927, p. 665
168. E.N., Jan. 19, 1911, p. 76; E.N.R., Apr. 21, 1932, p. 597; July 21, 1932, p. 90; Aug. 11, 1932, p. 178
169. E.N., Jan. 19, 1911, p. 76; Apr. 21, 1932, p. 597; July 21, 1932, p. 90; Aug. 11, 1932, p. 178
170. E.N.R., Mar. 25, 1937, p. 459

Chapter 11

171. A.F., July 1957, p. 6
172. E.N.R., July 12, 1956, p. 24
173. E.N.R., May 27, 1937, p. 787
174. E.N.R., Apr. 24, 1919, p. 813

Index